A GIFT FOR:

FROM:

The
Creation
Answer Book

HANK HANEGRAAFF

THOMAS NELSON
Since 1798

NASHVILLE DALLAS MEXICO CITY RIO DE JANEIRO

The Creation Answer Book

© 2012 Hank Hanegraaff

Published in Nashville, Tennessee, by Thomas Nelson. Thomas Nelson is a registered trademark of Thomas Nelson, Inc.

Author is represented by the literary agency of Alive Communications, Inc., 7680 Goddard Street, Suite 200, Colorado Springs, CO 80920, www.alivecommunications.com.

Thomas Nelson, Inc., titles may be purchased in bulk for educational, business, fund-raising, or sales promotional use. For information, please e-mail SpecialMarkets@ThomasNelson.com.

Italics in Scripture indicate the author's emphasis.

Unless otherwise noted, Scripture quotations are taken from the Holy Bible: New International Version®, NIV®. © 1973, 1978, 1984 by Biblica, Inc™. Used by permission of Zondervan. All rights reserved worldwide. www.zondervan.com

Scripture quotations marked NASB are from NEW AMERICAN STANDARD BIBLE®. © The Lockman Foundation 1960, 1962, 1963, 1968, 1971, 1972, 1973, 1975, 1977, 1995. Used by permission.

ISBN-13: 978-1-4003-1926-8

Printed in the China

12 13 14 15 16 WAI 6 5 4 3 2 1

To my daughter Faith—bright,
beautiful . . . beloved.

Contents

Creation and First Things

Creation and First Things: Advanced Reading

Creation and the Garden of Eden

Creation and the Flood

Creation and the Age Question

Creation and the Problem of Evil

Creation and Dinosaurs

Creation and Evolution

Creation and Evolution: Advanced Reading

Creation and Re-Creation

Foreword

*Since the creation of the world God's invisible
qualities—his eternal power and divine nature—have
been clearly seen, being understood from what has been
made, so that men are without excuse.*

—THE APOSTLE PAUL, ROMANS 1:20

Of one thing I am certain: people want answers.
And that is precisely what *The Creation
Answer Book* provides. Concise, well-researched
answers to *the* single most important apologetic
issue—that of origins. Indeed, how you view your
origins will ultimately determine how you live
your life. Think Lady Gaga. If Gaga, like Madonna
before her, is merely a material girl living in a
material world, her choices are not free—they
are fatalistically determined by such things as
brain chemistry and genetics. Conversely, if she is
created in the image of God, her life has eternal
meaning and significance.

First, *The Creation Answer Book* demonstrates
conclusively that the evolutionary hypothesis is no
longer tenable in an age of scientific enlightenment.

The Cambrian Explosion* (biology's Big Bang; words marked by an asterisk are defined in the glossary at the back of the book) has uprooted Darwin's Tree of Life*. And in the modern era of scientific enlightenment, it is untenable to suppose that something as seemingly simple as a human tear (which has no counterpart in the animal kingdom) evolved through unguided, purposeless processes. As you trek through these pages, it will become abundantly clear that "nothing" *could not* have created everything; that life *could not* have evolved from nonlife; and that random processes *could not* have produced morals.

Furthermore, *The Creation Answer Book* addresses questions that routinely stumble seekers and solidify skeptics in opposition to a biblical worldview: "If God created the universe, who created God? Was Eve tempted by a talking snake? If we can't see God, how do we know he exists? And what about dinosaurs? Did they die out 65 million years ago or live contemporaneously with humankind?"

Finally, *The Creation Answer Book* delves into one of the most controversial debates in all of Christianity—the debate about earth's age. In my previous books I left that issue on the backburner. In this book I tackle it head on! "Is this a young world after all? Or do indicators such as the speed of light and star life demonstrate that the universe is billions of years old? Are the creation days of Genesis literal, long, or literary? And what about animal suffering and death—is nature red in tooth and claw* a consequence of Adam's fall or part of God's 'very good' creation?"

The Creation Answer Book, however, is designed not only to answer questions but also to equip you to read the Bible in the sense in which it is intended. Thus, when you read that God walked with Adam and Eve in the cool of the day, you will not suppose that Moses intended to communicate that God has literal feet. Or when you read that God asked Adam, "Where are you?" you will not for a moment question God's omniscience. Moreover, if you are a believer, you will be prepared to give an answer for the hope that you have with gentleness and respect. And if you are not, you will have ample evidence to believe that the opening words of Scripture—*"In the beginning God"*—correspond to the age of scientific enlightenment in which you find yourself.

Hank Hanegraaff

Creation and
First Things

Who Made God?

First, unlike the universe, which according to modern science had a beginning, God is infinite and eternal. Thus, as an infinite, eternal being, God can logically be demonstrated to be the uncaused First Cause.

Furthermore, to suppose that because the universe had a cause, the cause of the universe must have had a cause simply leads to a logical dead end.

Finally, simple logic dictates that the universe is not merely an illusion. It did not spring out of nothing (nothing comes from nothing); and it has not eternally existed (the law of entropy* predicts that a universe that has eternally existed would have died of heat loss an "eternity ago"). The only plausible possibility that remains is that the universe was made by an unmade Cause greater than itself.

> "You are my witnesses," declares the LORD,
> "and my servant whom I have chosen,
> so that you may know and believe me
> and understand that I am he.
> Before me no god was formed,
> nor will there be one after me."
>
> ISAIAH 43:10

If We Can't See God, Can We Know He Exists?

It is not uncommon for skeptics to suppose that Christians are irrational to believe in a God they cannot "see." In reality, it is irrational for skeptics to suppose that what cannot be seen does not exist.

First, Christians and skeptics alike recognize black holes, electrons, laws of logic, and the force of gravity despite the fact that all these are unseen. Even the most ardent anti-supernaturalist recognizes the gravity of gravity.

Furthermore, "God's invisible qualities—his eternal power and divine nature—have been clearly seen, being understood from what has been made, so that men are without excuse" (Romans 1:20). Put another way, the order and the complexity of the universe testify to the existence of an uncaused First Cause.

Finally, Jesus is the image of the invisible God. As such, the incarnation of Christ is the supreme act of God's self-revelation. Thus, we can experience the power and presence of God in a way that is more fundamentally real than our perceptions of the physical world.

> Now we see but a poor reflection as in a mirror; then we shall see face to face. Now I know in part; then I shall know fully, even as I am fully known.
>
> 1 Corinthians 13:12

For further study, see Garrett J. DeWeese and J. P. Moreland, *Philosophy Made Slightly Less Difficult: A Beginner's Guide to Life's Big Questions* (Downers Grove, IL: InterVarsity Press, 2005).

How Many Explanations Are There for the Existence of Our Universe?

Philosophical naturalism*—the worldview that undergirds evolutionism—can provide only three explanations.

1. The universe is merely an illusion. This notion carries little weight in this age of scientific enlightenment.
2. The universe sprang from nothing. This proposition flies in the face of both the laws of cause and effect and of energy conservation (nothing comes from nothing; nothing ever could). Put another way, there simply are no "free lunches."
3. The universe eternally existed. The law of entropy* shatters this hypothesis: a universe that has eternally existed would have died of heat loss an "eternity ago."

There is, however, one other possibility. It is found in the first chapter of the first book of the

Bible: "In the beginning God created the heavens and the earth." In an age of empirical* science, no statement could be more certain, clear, or correct.

> For since the creation of the world God's invisible qualities—his eternal power and divine nature—have been clearly seen, being understood from what has been made, so that men are without excuse.
>
> ROMANS 1:20

For further study, see James W. Sire, *The Universe Next Door: A Basic Worldview Catalog*, 5th ed. (Downers Grove, IL: InterVarsity Press, 2009).

Could the Universe Emerge Out of Nothing?

When people face the compelling evidence that the universe began to exist at a definite point in time, a favorite fallback position is that it sprang into existence from nothing at all. This, however, stretches credulity beyond the breaking point.

First, simple logic dictates that nothing comes from nothing. "Nothing" is nonexistent and therefore lacks the *power to do*. Indeed, this "power to do" logically presupposes the existence of a thing that possesses that power.

Furthermore, something produced *by* nothing *from* nothing would, logically, have had to create itself. But if it created itself, it would have had to exist prior to its own creation, which means it must both exist and not exist at the same time and in the same way—an obvious contradiction and an utterly illogical conclusion. When the laws of logic are violated like this, reason and communication become meaningless.

Finally, in order for something to exist without

being the result of a prior cause, that something must be eternal (i.e., something that did not come into being, but has always existed). As such, the universe could not emerge out of nothing, but it can exist as an effect of an uncaused eternal First Cause—which is precisely what God is.

> Before the mountains were born
> or you brought forth the earth and the world,
> from everlasting to everlasting you are God.
>
> PSALM 90:2

Source (and for further study), see Norman L. Geisler, *Baker Encyclopedia of Christian Apologetics* (Grand Rapids: Baker, 1998), 399–401.

Can Chance Account for the Universe?

A stronaut Guy Gardner, who has seen earth from the perspective of the moon, points out that "the more we learn and see about the universe the more we come to realize that the most ideally suited place for life within the entire solar system is the planet we call home." In other words, life on earth must have been designed by a benevolent Creator rather than directed by blind chance.

First, consider the ideal temperatures on planet earth. If we were closer to the sun, we would fry. If we were farther away, we would freeze.

Furthermore, ocean tides, which are caused by the gravitational pull of the moon, play a crucial role in our survival. If the moon were significantly larger and therefore had a stronger gravitational pull, devastating tidal waves would submerge large areas of land. If the moon were smaller, tidal motion would cease, and the oceans would stagnate and die.

Finally, consider plain ol' tap water. The solid state of most substances is denser than their liquid state, but the opposite is true for water, which explains why ice floats rather than sinks. If water were like virtually any other liquid, it would freeze from the bottom up rather than from the top down, killing aquatic life, destroying the oxygen supply, and making earth uninhabitable.

From the temperatures to the tides to tap water and myriad other characteristics that we so easily take for granted, earth is an unparalleled planetary masterpiece. Like Handel's *Messiah* or da Vinci's *Last Supper*, our amazing planet should never be carelessly pawned off as the result of blind evolutionary processes.

In the beginning God created the heavens and the earth.

GENESIS 1:1

For further study, see R. C. Sproul, *Not a Chance: The Myth of Chance in Modern Science and Cosmology* (Grand Rapids: Baker, 1999).

Does the Fine-Tuning of the Universe Point to a Sovereign Creator?

One of the most astonishing discoveries of the twentieth century was that the universe is fine-tuned to support intelligent life. From the force of gravity to the balance of matter and antimatter, the universe is balanced, as it were, on the fine edge of a razor.

Consider the force of gravity. If it were stronger or weaker by just one part out of 10^{100} (that's a 1 with one hundred zeros after it!), the universe would not—could not—support intelligent life! To help grasp this number and not miss the gravity of gravity's fine-tuning, consider that the number of atoms in the entire observable universe is estimated to be only 10^{80}.

Furthermore, the fine-tuning of a force like gravity could not be a function of physical law. Why? Gravity could be stronger or weaker and still be gravity, so physical law does not dictate its precise strength. But were the force of gravity not fine-tuned to be the strength that it is, it could not support intelligent life.

Finally, the fine-tuning of the universe cannot reasonably be attributed to chance because of the infinitesimally small range of values involved. Chance is "infinitely" more likely to produce a life-prohibiting universe than a life-sustaining one.

The only plausible source of the fine-tuning of the universe is an external, transcendent, incalculably powerful and intelligent personal Mind, whom we call God.

Source (and for further study), see William Lane Craig, *Reasonable Faith: Christian Truth and Apologetics*, 3rd ed. (Wheaton, IL: Crossway, 2008).

Is Earth a Privileged Planet?

Some scientists speculate that the earth is an insignificant speck of soil aimlessly adrift in a meaningless universe. As documented by astronomer Guillermo Gonzalez and philosopher Jay W. Richards, however, evidence refutes the principle of mediocrity (Copernican principle), demonstrating instead that our earth is a singularly privileged planet designed for discovery.

First, the unique conditions necessary to support intelligent life turn out to provide the best overall conditions for scientific discovery. Examples abound. Earth is situated between two arms of a flattened spiral galaxy—the Milky Way—not too close to the core to be exposed to lethal radiation, comet collisions, or light pollution that would obscure observation of the distant universe; and not so far that a privileged planet could never form or where we would not observe different kinds of nearby stars. Moreover, the atmosphere of our privileged planet is both oxygen-rich for survival and transparent for discovery. The moon is the perfect size and distance from earth to stabilize rotation and to facilitate human habitability.

Not only so, the moon and sun's relative sizes and distances from earth provide perfect solar eclipses, which played a vital role in the development of modern science (e.g., determination of the nature of stars and confirmation of Einstein's general theory of relativity).

Furthermore, we live in the best overall age of the universe to do cosmology. In our time the cosmic background radiation left over from the Big Bang is readily observable, but it won't always be. Futhermore, this radiation confirms that the universe is not eternal but began in the finite past. Moreover, most of the astrophysical phenomena astronomers rely on to measure the universe were not observable earlier in the universe's development, and they will eventually fade (e.g., cosmic background radiation)—but neither could we have survived at earlier or later stages.

Finally, the setting of our privileged planet permits a stunning diversity of measurements, from the universe at large (cosmology) to the smallest of subatomic particles (quantum physics) to the middling size of earth and humans (geology and anthropology).

From habitability to discoverability, earth's

status in the universe is surely one of privilege. To reduce this to an accident of cosmic evolution is shortsighted; to recognize it as privileged, sublime.

It is the glory of God to conceal a matter;
to search out a matter is the glory of kings.

PROVERBS 25:2

Source (and for further study), see Guillermo Gonzalez and Jay W. Richards, *The Privileged Planet: How Our Place in the Cosmos Is Designed for Discovery* (Washington, DC: Regnery, 2004).

Did God Create Everything Out of Nothing?

The notion that God created everything from nothing has fallen on hard times. A surprising number of philosophers and theologians dogmatically contend that the doctrine of creation out of nothing (*ex nihilo*) has little scriptural support. Worse yet, leading Mormons overtly contend that matter has coexisted eternally with God. But the opening statement in Scripture—"In the beginning God created the heavens and the earth"—points to the truth that God created everything out of nothing. Indeed, only three options exist, and only one corresponds to reality.

First is the view that in the beginning nothing existed. Neither mass, nor energy, nor the Almighty. Nothing, nothing, simply nothing. Logic, however, screams that nothing comes from nothing!

Furthermore, there is the untenable notion that something existed, but that something was an impersonal potentiality out of which every potentiality—from protein molecules to personal mind—emerged. This idea, however, hardly advances the proverbial ball. As common sense tells us, every effect must have a cause equal to or greater than itself.

Finally, there is the scriptural contention, and of the three it alone makes sense: the universe was created by an uncaused First Cause greater than itself. Time, space, and the universe have not always existed, but God has always existed, and God's existence is the cause for the existence of all else that exists. While science demands that the universe had to have a beginning, nothing philosophically or scripturally demands that the *cause* of the universe had to have a beginning. As the writer of Hebrews aptly put it, "By faith [not blind faith but faith grounded in evidence] we understand that the universe was formed at God's command, so that what is seen was not made out of what was visible" (11:3).

Did God create everything out of nothing? Absolutely! The very first sentence of Scripture demands it. And an age of scientific enlightenment can abide nothing less.

By the word of the LORD were the heavens made,
their starry host by the breath of his mouth.

PSALM 33:6

For further study, see Paul Copan and William Lane Craig, *Creation Out of Nothing: A Biblical, Philosophical, and Scientific Exploration* (Grand Rapids: Baker Academic, 2004).

What Is the Doctrine of Creation Out of Nothing (Creatio ex Nihilo)?

T ry for a moment to think about nothing—no people, no plants, no water, no air, no matter, no energy, no time, no space, no God. Nothing. Simply nothing. The more we ponder such nothingness, the more clearly we realize that the notion of nothingness could not be further from the truth. To posit that God created everything out of nothing immediately resolves the issue. And that is precisely what the concept of creation *ex nihilo* does.

First and foremost, creation ex nihilo advances the doctrine that the eternally existing God created the universe out of nothing at all.

Furthermore, creation ex nihilo acknowledges that all that exists except God was created by God a finite time ago.

Finally, the doctrine of creation ex nihilo rules out the idea that such finite realities as space, time, and matter emanate from the essence of God: creation is neither divine nor a derivative of God's essence.

Simply put, the God of all that exists freely chose to speak, and the cosmos leaped into existence.

In the beginning was the Word, and the Word was with God, and the Word was God. He was with God in the beginning. Through him all things were made; without him nothing was made that has been made.

JOHN 1:1–3

Source (and for further study), see Paul Copan and William Lane Craig, *Creation Out of Nothing: A Biblical, Philosophical, and Scientific Exploration* (Grand Rapids: Baker Academic, 2004).

Can the Big Bang Be Harmonized with Genesis?

The Big Bang theory supposes that the universe began as an infinitely dense *singularity* and has since been expanding for billions of years. Though the Big Bang is not communicated in the Genesis account of creation, it lends scientific credibility to the scriptural contention that God created the universe *ex nihilo* (out of nothing).

First, like the Bible, the Big Bang theory hypothesizes that the universe had a beginning. This claim stands in stark opposition to the scientifically silly suggestion that the universe has eternally existed.

Furthermore, if the universe had a beginning, it had to have a cause. Indeed, the cause of all space, time, matter, and energy *must be nonspatial, nontemporal, immaterial, and unfathomably powerful and personal.* As such, the Big Bang flies in the face of the preposterous proposition that the universe sprang into existence out of nothing and lends credence to the Genesis contention of a Creator who spoke and the universe leaped into existence.

Finally, though evolutionists hold to the Big Bang, the Big Bang does not imply biological evolution. Big Bang cosmology answers questions about the origin of the space-time continuum, not questions about the origin of biological life on earth.

While we must not stake our faith on Big Bang cosmology, we can be absolutely confident that, as human understanding progresses, creation will continue to point to the One who spoke the universe into existence.

Source (and for further study), see William Lane Craig, *On Guard: Defending Your Faith with Reason and Precision* (Colorado Springs: David C. Cook, 2010).

Can God Create a Rock
So Heavy That He Cannot
Move It?

This question is a riddle that can leave most Christians looking like the proverbial deer in the headlights. At best, it challenges God's omnipotence. At worst, it undermines his very existence.

First, there is a problem with the premise of the question. While it is true that God can do anything that is consistent with his nature, it is absurd to suggest that he can do everything. For instance, God cannot lie (Hebrews 6:18). Neither can he be tempted (James 1:13), nor can he cease to exist (Psalm 102:25–27).

Furthermore, just as it is impossible to make a one-sided triangle, so it is impossible for God to make a rock too heavy for him to move. What an all-powerful God can create, he can obviously move. Put another way, God can do everything that is logically possible.

Finally, we should note that skeptics raise a wide variety of similar questions to undermine the

Christian view of God. Thus, it is crucial that we learn to question the question rather than readily assume that the question is valid.

> Do not answer a fool according to his folly,
> or you will be like him yourself.
> Answer a fool according to his folly,
> or he will be wise in his own eyes.

PROVERBS 26:4–5

What Is the Book of Nature?

When I refer to the book of nature*, people invariably wonder where they might get a copy. Thus, the question: What in the world *is* the book of nature?

First, *the book of nature* is a reference to general revelation. *Special revelation* reveals what God did to reconcile the world to himself (the Bible). *General revelation* reveals what God has shown in creation (the book of nature). As King David put it, "The heavens declare the glory of God; the skies proclaim the work of his hands," and "their voice goes out into all the earth, their *words* to the ends of the world" (Psalm 19:1, 4).

Furthermore, the book of nature reveals God's invisible qualities. The apostle Paul said exactly that: "For since the creation of the world God's invisible qualities—his eternal power and divine nature—have been clearly seen, being understood from what has been made, so that men are without excuse" (Romans 1:20).

Finally, the book of nature augments human reason through natural revelation. Indeed, it was failure to apply the explanatory power of natural revelation to the mysteries of the universe that trapped pagan thinkers in the intellectual cul-de-sac of their own making. Aristotle idealized the world in terms of how it ought to be. Augustine turned pages in the book of nature and discovered how it really was. Luther went so far as to say that "any potter has more knowledge of nature" than was written in the books of Aristotle.

Suffice it to say, Aristotle thought himself to be living in the quintessential age of invention and innovation—prejudice precluding the progress of science. In place of a rational God who orders the universe according to knowable principles, Socrates declared astronomical observations a "waste of time," and Plato persuaded his devotees to "leave the starry heavens alone." While obsessing on astrology, they left astronomy an unexplored domain. Likewise, they mastered the magic of alchemy* while remaining blithely ignorant of the majesty inherent in chemistry.

> *The heavens declare the glory of God;*
> *the skies proclaim the work of his hands.*
> *Day after day they pour forth speech;*

night after night they display knowledge.
There is no speech or language
where their voice is not heard.
Their voice goes out into all the earth,
their words to the ends of the world.

PSALM 19:1–4

For further study, see the epilogue in Hank Hanegraaff, *Has God Spoken? Proof of the Bible's Divine Inspiration* (Nashville: Thomas Nelson, 2011).

Can People of Science Be People of Scripture?

The realization that revelation is key to knowledge led medieval thinkers to crown theology "Queen of the Sciences." Peter Paul Rubens elegantly illustrated this truth in a seventeenth-century painting entitled *The Triumph of the Eucharist*. Standing in a chariot propelled by angelic beings was Theology. Walking alongside were Philosophy, a wise and grizzled veteran, and Science, a newcomer in the cosmic conversation. The point of the painting is this: theology is never absent philosophy and science, but philosophy and science absent theology inexorably leads to the blind ditch of ignorance. And Peter Paul Rubens was not the only one convinced of this truth.

First, Leonardo da Vinci, widely considered to be the real founder of modern science, was deeply committed to the scriptural truth—"In the beginning God created the heavens and the earth." Likewise, Robert Boyle, the father of modern chemistry as well as the greatest physical scientist of his or any other generation, was an able apologist for the Genesis account of origins.

Furthermore, Sir Isaac Newton—a remarkable intellect who developed calculus, discovered the law of gravity, and designed the first reflecting telescope—vigilantly defended the biblical account of creation. Louis Pasteur, best known for developing the concept of pasteurization and demolishing the concept of spontaneous generation*, underscored the power of revelation and undermined the evident absurdities of Darwinian evolution.

Finally, a veritable host of stars in the constellation of scientific discovery were people of science and people of God: Johannes Kepler (astronomy), Francis Bacon (scientific method), Blaise Pascal (mathematics and philosophy), Carolus Linnaeus (biological taxonomy), Gregor Mendel (genetics), Michael Faraday (electromagnetics), Joseph Lister (antiseptic surgery), Henrietta Swan Leavitt (astronomy), and Clara Swain (medicine).

These men and women were deeply committed to both Scripture and empirical* science. As such, they not only unpacked revelations within the book of nature* but were likewise illumined by the truth of biblical revelation.

Source (and for further study), see Henry M. Morris, *Men of Science, Men of God: Great Scientists of the Past Who Believed the Bible* (El Cajon, CA: Master Books, 1988).

Did God Reveal the Gospel in the Stars?

Over the years I have observed an alarming trend toward "magic apologetics." One of the more curious brands is the gospel in the stars (GIS). This theory asserts that God from the beginning wrote the unique message of the gospel in the signs of the zodiac. At first blush this idea may appear to have merit; closer examination, however, exposes it as completely untrue.

First, GIS compromises *sola Scriptura* (Latin for "by Scripture alone"). While sola Scriptura does not claim that the Bible is the sole source of revelation, it does maintain that Scripture is the sole repository of redemptive revelation. Moreover, nowhere in its pages does the Bible indicate that God has given us two infallible sources of special revelation—the gospel in the stars and the gospel in the Scriptures. Nor was GIS used by the prophets, the apostles, Jesus Christ, or the early church as an apologetic for the plan of salvation.

Furthermore, GIS confuses special revelation with general revelation. General revelation proclaims the glory of God through order and design* (Psalm 19:1). Special revelation is found in the "law of the LORD" set forth in the pages of Scripture (v. 7). From the lights we gain an unspoken knowledge of the Creator; the law is the schoolmaster that drives us to Christ. Common sense should suffice to tell us that the stars do not provide us with specific information about how to find salvation from our sin. Indeed, a common person looking at the night sky would be hard-pressed to see even a hint of God's plan of salvation in the zodiac.

Finally, GIS subverts the natural use of the stars, which God ordained, for a superstitious use, which he disdains. The natural uses of the stars are to "separate the day from the night," "serve as signs to mark seasons and days and years," and "give light on the earth" (Genesis 1:14–15). Stars can also be rightly used for varied purposes ranging from natural revelation to navigation. In light of the fact that GIS compromises sola Scriptura, it also confuses special revelation (revelation from the Bible) with general revelation (revelation in nature) and promotes superstition. Christians are to wisely reject it and instead master genuine apologetic arguments. Every Christian should be equipped to communicate that God created the universe, that Jesus Christ demonstrated that he is God through the immutable fact of the resurrection, and that the Bible is divine rather than merely human in origin.

For further study, see Hank Hanegraaff, "What Is Wrong with Astrology," *The Complete Bible Answer Book: Collector's Edition* (Nashville: J Countryman, 2009), 308–310; also Charles Strohmer, "Is There a Christian Zodiac, a Gospel in the Stars?" *Christian Research Journal*, vol. 22, no. 4 (2000).

Did Moses Write the Book of Genesis?

One complaint about the Genesis account of creation is that the first five books of the Bible, known collectively as the Pentateuch ("five-volumed book"), could not possibly have been written by Moses. Facts, however, say otherwise.

First, other biblical authors, including Joshua, Ezra, Daniel, and Paul, point squarely to Moses' authorship of the Pentateuch. Indeed, the Pentateuch itself claims Moses as its author. In Exodus, for example, we read that "the LORD said to Moses, 'Write down these words, for in accordance with these words I have made a covenant with you and with Israel.' Moses was there with the LORD forty days and forty nights without eating bread or drinking water. And he wrote on the tablets the words of the covenant—the Ten Commandments" (34:27–28).

Furthermore, Jesus also placed the divine stamp of authenticity on Mosaic authorship when he said, "If you believed Moses, you would believe me, for he wrote about me. But since you do not believe what he wrote, how are you going to believe what I say?" (John 5:46–47). Then, when Jesus spoke about marriage and divorce, he appealed directly to the words of Moses in Genesis 2 and added, "Moses permitted you to divorce your wives because your hearts were hard. But it was not this way from the beginning. I tell you that anyone who divorces his wife, except for marital unfaithfulness, and marries another woman commits adultery" (Matthew 19:8–9).

Finally, in addition to these testimonies from the Old and the New Testament, other significant factors support Mosaic authorship. For example, the name *Moses* itself provides corroborating evidence for the exodus. The name is both rooted in Egyptian tradition and fits well with the exodus era. In addition, the author of Genesis, Exodus, Leviticus, Numbers, and Deuteronomy was clearly an eyewitness familiar with Egyptian names, words, customs, plants, animals, and even geography. This knowledge would have been difficult to come by unless the author was, in fact, an eyewitness to the remarkable events of the Egyptian captivity, the exodus, the forty-year sojourn in the wilderness, and the final encampment before entering the Promised Land.

In sum, those who deny Moses' authorship of the first five books of the Old Testament seem influenced more by anti-supernatural bias than any biblical or historical evidence.

Philip found Nathanael and told him, "We have found the one Moses wrote about in the Law, and about whom the prophets also wrote—Jesus of Nazareth, the son of Joseph."

JOHN 1:45

For further study, see Hank Hanegraaff, *Has God Spoken? Proof of the Bible's Divine Inspiration* (Nashville: Thomas Nelson, 2011).

What Is the Genius of Genesis?

To say that Genesis is a literary masterpiece is to understate its elegance. With inspired brilliance, Moses interlaced a historical narrative with both symbolism and repetitive poetic structure, and he employed the powerful elements of story (characters, plot, tension, resolution) to set the foundation for the rest of redemptive revelation. The very chapter that references the fall also records the plan for restoration of fellowship—a plan that takes on definition with God's promise to make Abram a great nation through which "all peoples on earth will be blessed" (Genesis 12:3). As such, Abram's call is the divine antidote to Adam's fall. Yet this is but a glimpse of the author's genius. Genesis is forged so that its main message is as easy to remember as it is to recall.

To begin with, Genesis opens with a literary mnemonic (memory aid) by which we are daily reminded of God's creative prowess. The first six days outline a hierarchy of creation that culminates in humanity as its crowning jewel. On the seventh day, the Creator, in whom we ultimately find our Sabbath, rests. As such, the history of creation is remembered and recalled through its association with the continuous seven-day cycle of life.

Furthermore, the rest of Genesis is structured in a way that it may be remembered using our ten fingers. With one hand we recall primeval history: the accounts of the heavens and the earth (2:4–4:26), Adam (5:1–6:8), Noah (6:9–9:29), Noah's sons (10:1–11:9), and Shem, the father of the ancient Near East (11:10–26). With the other hand, we remember the accounts of Terah (father of Abraham, 11:27–25:11), Ishmael (25:12–18), Isaac (25:19–35:29), Esau (36:1–43), and Jacob, who is called Israel (37:2–50:26).

Finally, we should note that post-Gutenberg (printing press) we are primarily people of the printed page. We associate sound education with the capacity for reading and writing rather than memorizing and reciting. Not so the ancients. In their predominantly oral culture, people practiced the principles of memory. As such, Genesis contains many Hebrew symmetries, parallelisms, and sevenfold patterns. An example of a symmetrical pattern is found in the record of humanity's first sin

(Genesis 2:4–3:24: the creation of man and woman; temptation by the serpent; sin in the center; and punishment of the serpent, woman, and man). The account of the seven-day creation (Genesis 1:3–27) is a sevenfold pattern that contains a three-way parallel structure: Days 1 and 4, light/luminaries; Days 2 and 5, sky and sea/sea and sky creatures; Days 3 and 6, land/land creatures; and Day 7, the Sabbath.

From its seven-day opening through its tenfold pattern, Genesis serves as a unified and memorable prologue to the whole of redemptive history.

For further study, see David A. Dorsey, *The Literary Structure of the Old Testament: A Commentary on Genesis–Malachi* (Grand Rapids: Baker, 2004).

What Are the Three Great Apologetic Issues?

There are three great apologetic issues—origins, resurrection, and biblical authority. Chief among these is the issue of origins.

First, how you view your origins will determine how you live your life. If you suppose you are a function of random processes, you will live life by a wholly different standard than if you know you are created in the image of God and accountable to him.

Furthermore, in a Christian worldview, the transcendent God who laid the foundations of the earth condescended to cloak himself in human flesh. The God-man Jesus not only died so that we might live, but he demonstrated that he was the Creator of life by raising himself from the dead. As such, Christ does not stand in a line of peers with Abraham, Buddha, or Confucius.

Finally, we can know with certainty that the Book, beginning with the words "In the beginning God" is the infallible repository of redemptive revelation. As Christians we do not accept this truth by blind faith but rather by faith rooted in fact.

> But in your hearts set apart Christ as Lord. Always
> be prepared to give an answer to everyone who asks
> you to give the reason for the hope that you have.
> But do this with gentleness and respect.
>
> 1 PETER 3:15

For further study, see these Hank Hanegraaff titles: *The FACE That Demonstrates the Farce of Evolution* (Nashville: Thomas Nelson, 2001); *Resurrection* (Nashville: Thomas Nelson, 2002); and *Has God Spoken? Proof of the Bible's Divine Inspiration* (Nashville: Thomas Nelson, 2011).

Creation and
First Things:
Advanced Reading

Is the Fine-Tuning of the Universe Negated by the Notion of a Multiverse?

Theoretical physicist Stephen Hawking is infamous for supposing that the existence of a vast number of universes—a multiverse—explains the fine-tuning of our universe. In other words, Hawking asserts that, given a sufficient number of random universes, one of them is bound to have the necessary conditions to support not only intelligent life but also the proposition that blind chance can account for fine-tuning. In reality, the multiverse proposition is an utterly desperate attempt to account for an unfathomably fine-tuned universe.

First, we should note that there is not a shred of evidence to support the existence of a physical universe other than our own, much less a virtually infinite number of such universes. Hawking's hope—and the hope of multitudes who share such fanciful presuppositions—is based on theological and theoretical pining rather than scientific discovery.

Furthermore, in addition to undermining science, the multiverse hypothesis throws plain old common sense under the bus. Imagine trying to convict a murderer in a multiverse where material evidence has been sacrificed on the altar of philosophically improbable propositions. In such a multiverse, stories of revolvers materializing out of thin air might well be as credible as eyewitness testimony.

Finally, eminent theoretical physicist—and Hawking colleague—Roger Penrose, though himself agnostic, has rightly concluded that the multiple universe hypothesis is *both "impotent" and "misconceived."* The Big Bang turns out to be an event of almost infinitesimal probability, and the probabilistic resources provided by a multiverse turn out to be insufficiently rich. As such, the most plausible explanation for the fine-tuning of our universe is yet, "In the beginning God created the heavens and the earth."

Source (and for further study), see William Lane Craig, *Reasonable Faith: Christian Truth and Apologetics*, 3rd ed. (Wheaton, IL: Crossway, 2008).

Why Is Creation *ex Nihilo* Theologically Significant?

Although creation *ex nihilo* has been compromised, confused, and contradicted, nothing could be more certain, correct, or theologically significant. Why? Because the doctrine of creation out of nothing (*ex nihilo*) implies God's necessary existence, underscores his divine freedom, and exhibits his divine omnipotence.

First, creation out of nothing bolsters the notion of God's necessary existence as the only Being who cannot *not* be. As such, the church fathers described the Father of creation as uncreated and unbegotten, in contrast to all else that was created and begotten. Put another way, all that exists, except God himself, is necessarily contingent on and grounded in the creative decisions and will of God.

Furthermore, creation ex nihilo also calls attention to God's freedom to create or to act otherwise. As such, the cosmos and all that is in it was neither mandatory nor a mishap. God freely chose to create humans and a habitat distinct from himself.

Finally, the doctrine of creation out of nothing underscores the reality that God alone is omnipotent. A God who creates out of eternally existing matter is less than the omnipotent Sovereign of the universe who spoke, and all that is leaped into existence.

Is creation ex nihilo theologically significant? Indeed! It makes all the difference in the world.

> For this is what the LORD says—
>   he who created the heavens,
>   he is God;
>   he who fashioned and made the earth,
>   he founded it;
>   he did not create it to be empty,
>   but formed it to be inhabited—
>   he says:
>   "I am the LORD,
>   and there is no other."
>
> ISAIAH 45:18

Source (and for further study), see Paul Copan and William Lane Craig, *Creation Out of Nothing: A Biblical, Philosophical, and Scientific Exploration* (Grand Rapids: Baker Academic, 2004).

Does the Bible Teach That God Created a Flat Earth?

I n *A History of the Warfare of Science with Theology in Christendom*, Andrew Dickson White, president and founder of Cornell University, decries the regrettable reality that two hundred years after Ferdinand Magellan had empirically* proven that the earth was round (1519), Christian fundamentalists persisted in perpetuating flat earth mythology. This, however, hardly corresponds to reality.

First, the moniker "flat earth" is propaganda. Some reading these words may well remember exactly where they were when they first heard the tale of Columbus' raw courage in face of mutinous sailors in mortal terror of sailing over the edge of a flat earth. What we are largely unfamiliar with is that far from fanatics clinging to flat earth mythology, unanimous scholarly opinion from Augustine to Aquinas pronounced the earth spherical.

Furthermore, the notion that evolutionary man is more enlightened than early man is but the vestigial* prejudice of Darwinist dogma. While modern man has an accumulation of knowledge that has produced innovations such as iPods, the genius that produced pyramids did not suddenly become benighted* when gazing at an ancient sky. It doesn't take a rocket scientist to recognize that a lunar eclipse is not produced by a flat earth. The problem is that postmodern man does science well but has become increasingly unsophisticated in the art and science of biblical interpretation. As such, poetic language is being forced to "walk on all fours." Isaiah 40:22 immediately springs to mind: "It is He who sits above the circle of the earth, and its inhabitants are like grasshoppers, who stretches out the heavens like a curtain and spreads them out like a tent to dwell in" (NASB). Some employ this passage as evidence for sphericity ("circle of the earth"); others for Big Bang cosmology ("spreads out"); still others for the notion that the Bible teaches a flat earth ("tent"). As Augustine made clear, however, Isaiah's language is self-evidently metaphorical. To read Isaiah in any other way would lead to the further absurdity that God lives in a physical mansion and drives an exotic chariot.

Finally, the notion that the enlightenment of the Greco-Roman world was divorced from the Renaissance by the deliberate obscurity of medieval churchmen of the Dark Ages is revisionist history at its worst. The millennium that encompassed Greek and Roman history is more correctly characterized by irrational superstition than rational thinking—Greco-Roman thought shackled to the irrational assumption of an eternal universe ministered by moody gods. Little wonder, then, that almost a thousand years after Aristotle, aristocrats spoon-fed at the table of Greek enlightenment dwelled in drafty domains never dreaming of a coming Christian era in which the invention of chimneys, clocks, and capitalism would revolutionize Western civilization.

In sum, flat earth mentality is more appropriately attached to evolutionists like Darwin who claimed that man can attain a higher importance in whatever he takes up than can woman. Or Enlightenment philosopher David Hume, who rendered blacks "naturally inferior" to whites.

For further study, see Rodney Stark, *The Victory of Reason: How Christianity Led to Freedom, Capitalism, and Western Success* (New York: Random House, 2005).

What About the Gap Theory?

The *Gap theory*—or *Ruin/Reconstruction theory*—arose with the dawn of modern science as an attempt to reconcile the geological age of the cosmos with the Genesis account of creation. Traceable to nineteenth-century Scottish theologian Thomas Chalmers and the twentieth-century *Scofield Reference Bible*, the Gap theory suggests there was a vast geological age of ruin between the first two verses of Genesis. Despite persistent popularity, three concerns continue to haunt the theory.

First, this *geological gap* seems wholly driven by scientific discovery rather than scriptural discernment. Million-year-old fossils discovered by paleontologists are supposed to be the remains of plants and animals destroyed in a flood that happened before Adam lived. Popularly referred to as *Lucifer's flood*, this ruin of the earth was God's judgment of the devil and demons. As the argument goes, the destruction happened between the events reported in Genesis 1:1 and Genesis 1:2, and this destruction was followed by a seven-day reconstruction period that is recounted beginning with Genesis 1:3.

Furthermore, this gap seems wholly made up. There is no mention of a gap in either the immediate context or anywhere else in Scripture. As such, it is a makeshift improvisation designed to resolve a perceived conflict between science and Scripture: science seems to demonstrate that physical evil preceded Adam, and Scripture documents that physical and personal evils stem from Adam's fall (Genesis 3:17–18; Romans 5:12–21; 8:19–21; 1 Corinthians 15:20–26; see the entries "Is Animal Suffering a Consequence of Adam's Sin?" on page 124 and "Could Carnivores and Catastrophes Have Existed Prior to the Fall?" on page 127).

Finally, there is the matter of a text taken out of context and turning into a *pretext* (false premise). A classic example is found in the notes of the *Scofield Reference Bible*. The initial pretext is that *create* and *made* have distinctively different meanings— *create* (*bara*) meaning "to make out of nothing" and *made* (*'asah*) meaning "to remake or reconstruct something previously created." Thus, the notion arose that the earth was *created* in 1:1, *made void waste and empty by judgment* between 1:1 and 1:2, and *reconstructed* thereafter. Scripture, however, uses the words *create* and *made* interchangeably. The single example of Genesis 1:26–27 is sufficient evidence: here God says, "Let us *make* (*'asah*) man in our image. . . . So God *created* (*bara*) man in his own image."

A little digging uncovers a host of similar pretexts used to support a *geological* gap that is ad hoc. Suffice it to say, the Bible does not provide a *chronology* of creation. More important to God's people, however, Scripture's creation account does provide an order of creation, a *hierarchy*, culminating in the crowning jewel of creation—humankind.

In the beginning God created the heavens and the earth. Now the earth was formless and empty, darkness was over the surface of the deep, and the Spirit of God was hovering over the waters. And God said, "Let there be light," and there was light.

GENESIS 1:1–3

For further study, see A. F. Johnson, "Gap Theory" in *Evangelical Dictionary of Theology*, ed. Walter E. Elwell (Grand Rapids: Baker, 1984), 439.

Creation and
the Garden
of Eden

Did Adam and Eve
Really Exist?

A growing number of Christian thinkers say no. Francis Collins, theistic evolutionist and founder of the BioLogos Foundation and tapped by President Obama to be Director of the National Institutes of Health (NIH), is among them. In his view, science dictates that modern humans emerged from primates a hundred thousand years ago in a population numbering some ten thousand, not two. Not only so, but according to BioLogos biblical expert Peter Enns, "The Bible itself invites a symbolic reading" of the account of the first man and woman. Thus the question: Did Adam and Eve exist, or are they merely allegorical?

First, while Collins and company staunchly defend Darwinian evolution, their views hardly correspond to reality. Darwin hung his hopes on hundreds of thousands of transitional forms* leading to the fossils of the Cambrian Explosion*. In actuality, the poverty of the fossil record has been an embarrassment. Virtually all known body plans appear abruptly in the Cambrian. To put it bluntly, the Cambrian radiation* vaporized the Darwinian Tree of Life*.

Furthermore, Scripture plainly opposes the collective rhetoric of theistic evolutionists who deny the reality of a historical Adam and Eve. Paul made it crystal clear: "From *one* man [God] made every nation of men, that they should inhabit the whole earth" (Acts 17:26). Indeed, sacred Scripture, in concert with sound science, makes plain that *kinds* reproduce "according to their kinds" (Genesis 1:25). Moreover, had the first Adam not fallen into a life of perpetual sin terminated by death, there would have been no need for God to send a "Second Adam" (Jesus). Paul was emphatic: "Since death came through a man, the resurrection of the dead comes also through a man. For as in Adam all die, so in Christ all will be made alive" (1 Corinthians 15:21–22).

Finally, our Savior's words should cast a pall on all Adam and Eve deniers: "At the *beginning* the Creator 'made them male and female'" (Matthew 19:4). Lest one be tempted to allegorize the words of our Lord, it is instructive to note that Jesus further affirmed a historical Adam and Eve when he referred to the murder of their son Abel (Matthew 23:35). Not only so, but Luke, writing to a primarily Gentile audience, extends his genealogy past Abraham to the first Adam, thus highlighting Christ, the Second Adam, as the Savior of all humanity. Should that prove insufficient, Chronicles provides a historical record from Adam to the Exile. Likewise, starting at Genesis 5:1, Moses gives "the written account of Adam's line."

Of one thing we can be absolutely certain: though Genesis is historical narrative interlaced with Jewish poetry, it is hardly an allegory.

What Is the Tree of Life?

The wisest man who ever lived likened *wisdom*, "the fruit of the righteous" and "longing fulfilled," to the proverbial "Tree of Life" (Proverbs 3:18; 11:30; 13:12). Such exemplars of beauty and blessing, however, find ultimate root in two gardens, with Golgotha in between.

In the beginning, the Tree of Life stood as the centerpiece of the garden of Eden. When Adam and Eve ate the forbidden fruit, the tree remained a memorial to Paradise lost: God "placed on the east side of the garden of Eden cherubim and a flaming sword flashing back and forth to guard the way to the tree of life" (Genesis 3:24).

Furthermore, standing on the other side of history, the Tree of Life is rooted in an eternal garden, this time, a memorial to Paradise regained. The angel of the apocalypse showed John, the apostle of the apocalypse, "the river of the water of life, as clear as crystal, flowing from the throne of God and of the Lamb down the middle of the great street of the city. On each side of the river stood the *tree of life*, bearing twelve crops of fruit, yielding its fruit every month. And the leaves of the tree are for the healing of the nations. No longer will there be any curse" (Revelation 22:1–3). "To him who overcomes," said Jesus, "I will give the right to eat from the *tree of life*, which is in the paradise of God" (2:7).

Finally, the Tree of Life stands on Golgotha's hill as the fulcrum of history. On it, Jesus stretches one hand toward the Garden of Eden, the other toward the eternal Garden. The immortality the first Adam could no longer reach, the Second Adam touched in his place. Thus, Jesus vanquished the power of evil, giving ultimate victory to the knowledge of good.

> How blessed is the man who finds wisdom
> And the man who gains understanding.
> For her profit is better than the profit of silver
> And her gain better than fine gold.
> She is more precious than jewels;
> And nothing you desire compares with her.
> Long life is in her right hand;
> In her left hand are riches and honor.
> Her ways are pleasant ways,
> And all her paths are peace.
> She is a tree of life to those who take hold of her,
> And happy are all who hold her fast.

PROVERBS 3:13–18 NASB

What Is the Tree of the Knowledge of Good and Evil?

The Tree of the Knowledge of Good and Evil is a prime example of physical reality highlighting spiritual truth (see John 3:12).

To begin with, God placed the tree in the middle of Paradise as a test; Satan perverted it as a temptation. In eating from the tree about which God said, "When you eat of it you will surely die," Adam challenged God as arbiter of good and evil (Genesis 2:17).

Furthermore, the Tree of the Knowledge of Good and Evil is the symbol of choice—between obedience and disobedience, between majestic revelation (God's truth) and moral relativism (my truth).

Finally, though Adam sinned, God responded with mercy. He drove Adam from the garden so that he no longer had access to the Tree of Life. To do otherwise would have doomed humanity to an endless physical existence in a fallen spiritual condition—which is precisely what hell is. Hell, however, can be averted. And Paradise regained.

As Adam's descendants we can yet embrace the Tree of Life where the personification of all that is good paid the ultimate price to take all that is evil upon himself.

> The LORD God made all kinds of trees grow out of the ground—trees that were pleasing to the eye and good for food. In the middle of the garden were the tree of life and the tree of the knowledge of good and evil.
>
> GENESIS 2:9

Why Does Genesis Portray Satan As an Ancient Serpent?

The Bible describes Satan as a twisted serpent whom God alone can crush. The metaphor is pregnant with meaning.

First, the biblical imagery of the twisted serpent is a masterful portrayal of Satan's fall. Once he held a privileged position in God's kingdom, but now he is even lower than the livestock. They have legs. He, however, crawls on his belly and eats dust. Those who know the Old Testament immediately recognize the metaphor. Micah, like Moses, uses the imagery of a serpent to depict the nations seeking to thwart the purposes of God: "They will lick dust like a snake, like creatures that crawl on the ground" (7:17).

Furthermore, Moses' original audience was intimately acquainted with the imagery of the serpent. During their sojourn in the desert, the lethal venom of fiery serpents had been emblazoned upon the tablet of their consciousness. Likewise, they would ever remember the bronzed serpent Moses lifted up in the Arabah[*] (Numbers 21:4–9). As the bronzed serpent was an image without venom, so the image of the invisible God who came in the appearance of sinful flesh was without sin. Thus, the serpent was not just emblematic of seduction, but the exemplar of a Savior who would die so that we may live (John 3:14–15).

Finally, the imagery of a serpent is employed by Scripture as literary subversion* of pagan myth. Though Baal, who according to the Ugaritic texts (pre-Jewish texts of the ancient Canaanites), rides the clouds and smites the primordial seven-headed serpent lying coiled in the chaotic waters, in reality it is Yahweh, Creator of heaven and earth, who alone can crush the serpent's head (Genesis 3:15). In sum, the Bible uses the pagan imagery of a serpent as a powerful theological apologetic by which to undress gods of wood and stone that are impotent to save.

In that day,
the LORD will punish with his sword,
his fierce, great and powerful sword,
Leviathan the gliding serpent,
Leviathan the coiling serpent;
he will slay the monster of the sea.

ISAIAH 27:1

Was Eve Deceived by a Talking Snake?

First, to read Scripture *literally* is to read it as *literature*. This means that we are to interpret the Word of God just as we interpret other forms of communication—in the most obvious and natural sense. Thus when Moses uses the symbolism of a snake, we do violence to his intentions if we interpret him in a woodenly literal fashion.

Furthermore, a literalistic interpretation does as much violence to the biblical text as a spiritualized interpretation that empties the text of all objective meaning. If the historical Adam and Eve did not eat the forbidden fruit and descend into a life of habitual sin, resulting in death, there would be no need for redemption. If, on the other hand, Genesis were reduced to an allegory conveying merely abstract ideas about temptation, sin, and redemption, devoid of any correlation with actual events in history, the very foundation of Christianity would be destroyed.

Finally, when the prophet Moses describes Satan as an ancient serpent and the apostle John describes him as an ancient dragon, they do not intend to tell us what Satan *looks* like. They want to teach us what Satan *is* like. (Dragons, after all, are the stuff of mythology, not theology.) If we think of Satan as either a slithering snake or a fire-breathing dragon, we would not only misunderstand the nature of fallen angels, but we might also suppose that Jesus triumphed over the work of the devil by stepping on the head of a snake (Genesis 3:15) rather than through his passion on the cross (Colossians 2:15).

In short, Eve was not deceived by a talking snake. Rather Moses used the symbol of a snake to communicate the wiles of the evil one who deceived Eve through mind-to-mind communication—precisely as he seeks to deceive you and me today.

> Now the serpent was more crafty than any of the wild animals the LORD God had made. He said to the woman, "Did God really say, 'You must not eat from any tree in the garden'?"
>
> GENESIS 3:1

For further study, see Hank Hanegraaff, *The Apocalypse Code: Find Out What the Bible Really Says About the End Times and Why It Matters Today* (Nashville: Thomas Nelson, 2008).

Do Genesis 1 and 2 Contradict Each Other?

It is not uncommon today to hear skeptical professors on university campuses assert that the Bible cannot be the infallible repository of redemptive revelation because the first two chapters are contradictory. How? In chapter 1, the creation of plants precedes the creation of man and animals, whereas in chapter 2, the creation of man precedes the creation of plants and animals.

First, it is highly unlikely that an author would contradict himself within the span of several sentences. Moreover, given the sophistication of the literary genres employed in Genesis, one is immediately alerted to a deeper purpose within the narrative. Rather than mining Genesis for all its wealth, fundamentalist fervor seems bent on forcing the language into a literalistic labyrinth from which nothing but nonsense can emerge.

Furthermore, even a cursory reading of Genesis 1 and 2 should be enough to discern that the author has a different purpose in one than in the other. Chapter 1 presents a three-level *hierarchy* of God's creative prowess, memorably associated with days of the week. In contrast, chapter 2 focuses on the crowning jewels of God's creation—man and woman—who are designed to be in right relationship with both creation and the Creator.

Finally, we must always remember that the language of Scripture is a heavenly condescension so that we finite human beings can know something of the nature and purposes of our infinite God. Readers concerned with a chronology of creation need look no further than God's revelation in the book of nature*. Indeed, those who tenaciously follow evidence wherever it leads will read both the book of Scripture and the book of science with an open mind.

> No shrub of the field had yet appeared on the earth and no plant of the field had yet sprung up, for the LORD God had not sent rain on the earth and there was no man to work the ground, but streams came up from the earth and watered the whole surface of the ground—the LORD God formed the man from the dust of the ground and breathed into his nostrils the breath of life, and the man became a living being.

GENESIS 2:5–7

What Does It Mean to Be Created in the Image and Likeness of God?

Genesis 1:26 asserts that humankind is made in the *imago Dei* (quite literally "the image of God"). While we do not share the *incommunicable* attributes of God—his omnipotence, omniscience, or omnipresence—we do share in a finite and imperfect way such *communicable attributes* as spirituality (John 4:24), rationality (Colossians 3:10), and morality (Ephesians 4:24). As such, we, the crowning jewels of God's creation, are reflections rather than reproductions of God.

First, we should note that, despite the fall, humanity continues to *reflect* the image of God. This reflection of his divine image in us has been blemished yet not obliterated. James, the half brother of Jesus, affirmed this truth when he wrote that men are "made in God's likeness" (James 3:9).

Furthermore, through sanctification, God is *renewing* an image that as yet is blemished and broken. We have taken off the "old self with its practices and have put on the new self, which is being *renewed* in knowledge in the image of its Creator" (Colossians 3:9–10).

Finally, it is liberating for believers to recognize that God will completely *restore* the imago Dei in fallen humanity. Paul wrote, "Just as we have borne the likeness of the earthly man, so shall we bear the likeness of the man from heaven" (1 Corinthians 15:49). We have the certain promise that one day the imago Dei now reflected in the crowning jewels of God's creation will be completely restored in Christ, who "is the image of the invisible God, the firstborn over all creation" (Colossians 1:15).

Though we will never be like God who has "life in himself" (John 5:26), we will through Christ "participate in the divine nature and escape the corruption in the world caused by evil desires" (2 Peter 1:4).

> So God created man in his own image,
> in the image of God he created him;
> male and female he created them.
>
> GENESIS 1:27

For further study, see Hank Hanegraaff, *Christianity in Crisis: 21st Century* (Nashville: Thomas Nelson, 2009).

Why Did God Accept Abel's Offering and Reject Cain's?

First, we must not suppose that Abel's offering was accepted because it was a blood sacrifice and Cain's rejected because it was bloodless. Each gave according to his vocation, but Abel gave the firstborn, the fattest of his flock, while Cain gave fruit that was not the firstfruits. As such, at issue was a disposition of the heart.

Furthermore, Cain's disposition is evident in his angry and downcast countenance. Sin reigned in him with such vulgar vengeance that he embodied the character and craftiness of his father the devil—"a murderer from the beginning" (John 8:44). Cain invited an unsuspecting Abel to the field and there murdered him with malice and forethought.

Finally, Cain's evil disposition is evidenced in the complaint regarding the severity of his punishment. He was bothered not by the content of his sin but by its consequences. It was not penitence but punishment that caused Cain to plead for protection from Adam's posterity.

Cain's offering in Genesis highlights the reality that man looks on the outward appearance, but the eyes of God pierce externalities and perceive the intent of the heart.

The LORD looked with favor on Abel and his offering, but on Cain and his offering he did not look with favor. So Cain was very angry, and his face was downcast.

GENESIS 4:4–5

Where Did Cain Get His Wife?

One of the most common objections to the Genesis account of creation is the reference to Cain's wife in Genesis 4:17. Unless God supernaturally created a wife for Cain as he had for Adam, he would have had to engage in incest with one of his sisters.

First, we should note that Adam lived almost a thousand years (Genesis 5:5) and fulfilled God's charge to "be fruitful and increase in number" (Genesis 1:28). Thus, while Scripture does not tell us where Cain got his wife, the logical implication is that he married either a sibling or a niece.

Furthermore, since genetic imperfections accumulated gradually over time, there was no prohibition against incest in the earlier stages of human civilization. God issued the Levitical law against incestuous relationships at the time of Moses, hundreds of years after Cain. So familial relationships were preserved and birth defects were prevented (see Leviticus 18:6, 9).

Finally, the speculation that God may have

created a wife for Cain as he had for Adam is completely ad hoc. The consistent teaching of Scripture is that "from one man [God] made every nation of men, that they should inhabit the whole earth; and he determined the times set for them and the exact places where they should live. God did this so that men would seek him and perhaps reach out for him and find him, though he is not far from each one of us" (Acts 17:26–27).

> Cain lay with his wife, and she became pregnant
> and gave birth to Enoch.
>
> GENESIS 4:17

For further study, see Gleason Archer, *Encyclopedia of Bible Difficulties* (Grand Rapids: Zondervan, 1982).

Creation and the Garden of Eden: Advanced Reading

What Are Essential Elements of the Two Creation Accounts?

While Bible-believing Christians may disagree on how to interpret Genesis 1 and 2, they all hold to the following three essentials.

First, they believe the Bible, including Genesis, to be the infallible repository of redemptive revelation. They may differ on interpretation but never on inspiration.

Furthermore, they are committed to the truth that God created the universe out of nothing at all (*creatio ex nihilo*). Moreover, they hold to horizontal changes within living kinds (microevolution*, as when bacteria becomes resistant to antibiotics), but never vertical changes from one kind to another kind (macroevolution*).

Finally, they hold Adam and Eve to be special creations made in the image of God rather than fictional exemplars of primitive humanity. As such, they deny naturalistic paradigms—including a version of theistic evolution in which God is said to have employed purely natural processes to produce the first humans.

If such naturalistic paradigms are true, Genesis is at best an allegory and at worst a farce. And if Genesis is either an allegory or a farce, the rest of the Bible is utterly irrelevant. In other words, if Adam and Eve did not eat the forbidden fruit and fall into a life of perpetual sin terminated by death, there is no need for redemption.

> The earth is the LORD's, and everything in it,
> the world, and all who live in it;
> for he founded it upon the seas
> and established it upon the waters.

PSALM 24:1–2

For further study, see *The Genesis Debate: Three Views on the Days of Creation*, ed. David G. Hagopian (Mission Viejo, CA: Crux Press, 2001).

Who Is the "Us" in Genesis 1:26?

Commentators variously refer to the *"us"* in Genesis 1:26 as angels, a plural of *majesty*, or a divine plurality. Which interpretation is correct?

First, though the Bible pictures God as surrounded by an angelic host who worship him and carry out his commands, there is no biblical basis for suggesting that angels took part in the creation of humankind or that humankind was created in the image of angels. Indeed, humans are said to be created "in the image of God" (v. 27).

Furthermore, there is no biblical precedent for the notion that Genesis 1:26 employs the first-person-plural-pronoun *us* to refer to God the Father in a fuller, more majestic sense—i.e., a plural of *majesty*.

Finally, there *is* warrant in the immediate and broader contexts to support the idea that the plural pronoun *us* refers to the divine plurality of the Trinity. In the immediate context, the man is said to be created in plurality (male and female), thus forging the finite relationship of a man and a woman in the image of infinite relationships within the Godhead. Likewise, the broader context of Scripture undergirds the *us* as one God revealed in three Persons, eternally distinct.

Then God said, "Let us make man in our image, in our likeness, and let them rule over the fish of the sea and the birds of the air, over the livestock, over all the earth, and over all the creatures that move along the ground."

GENESIS 1:26

If God Is One, Why Does the Bible Refer to Him in the Plural?

How could the Israelites be fiercely mono-theistic and yet refer to their God using the plural word *Elohim*?

First, this cannot be explained away as a "royal plural" or a "plural of *majesty*." Hebrew Scriptures never used a first-person plural to refer to any speaker other than God himself (e.g., Genesis 1:26).

Furthermore, while the Bible from Genesis to Revelation reveals that God is one in nature or essence (Deuteronomy 6:4; Isaiah 43:10; Ephesians 4:6), it also reveals that this one God eternally exists in three distinct Persons: Father, Son, and Holy Spirit (1 Corinthians 8:6; Hebrews 1:8; Acts 5:3–4). Thus, the plural ending of *Elohim* points to a plurality of Persons, not to a plurality of gods.

Finally, although *Elohim* is certainly *sugges-tive* of the Trinity, the word alone is not *sufficient* to prove the Trinity. Instead of relying on a singu-lar grammatical construction, Christians must be equipped to demonstrate that the one God revealed in Scripture exists in three Persons who are eternally distinct.

> Hear, O Israel: The LORD our God, the LORD is one.
>
> DEUTERONOMY 6:4

For further study, see Robert Letham, *The Holy Trinity: In Scripture, History, Theology, and Worship* (Phillipsburg, NJ: P&R Publishing, 2004).

What Is the Significance of the Proto-Evangel (Genesis 3:15)?

Eve was tempted to do the unthinkable—to take the place of God as arbiter of good and evil. Satan tempted, and the woman tasted temptation's ripened fruit. God authored the potential for evil by providing the woman with choice; the woman actualized evil by exercising that choice. Yet the very passage that references humanity's fall provides—in embryo—the prophetic antidote.

First, we should note that *proto-evangel* is a compound word meaning "first (*proto*) gospel (*evangel*)." As such, the proto-evangel offers the first note of hope and redemption following humanity's fall into perpetual sin terminated by death.

Furthermore, we experience the gospel in embryo as it grows to full maturity by reading the rest of the story. From Adam's rebellion to Abraham's Royal Seed, Scripture chronicles God's unfolding plan of redemption: the serpent would strike the Savior's heel, and the Savior would forever crush its head.

Finally, to pray, "Thy kingdom come," is to remember afresh that Christ has already won the war, but the reality of his reign is not yet fully realized. At present we are sandwiched between the triumph of the cross and the termination of time—between D-day and V-day. D-day was Christ's First Coming when Satan was decisively defeated. V-day is the Second Coming when Paradise lost will be Paradise regained.

History is hurtling toward a glorious and climactic end when the kingdoms of this world will become the kingdom of our Lord. A day in which we, like Adam and Eve, will once again walk with God in the cool of the day.

> *"And I will put enmity*
> *between you and the woman,*
> *and between your offspring and hers;*
> *he will crush your head,*
> *and you will strike his heel."*

Genesis 3:15

Creation and
the Flood

Why Are Corrupted Flood Accounts Significant?

With all their failings, corrupted accounts of the Flood—such as the Epic of Gilgamesh, the Epic of Atrahasis, and the even older Eridu Genesis—are yet a significant treasure.

First, similarities between the Genesis Flood account and corrupted Flood accounts are best explained by a "common inheritance." Put another way, they all derive from the same source—an actual event. It stands to reason that, as descendants of Noah drifted from God and from one another, human embellishments and interpretations would be imported into the actual historical event.

Furthermore, the existence of corrupted Flood accounts underscores the reality of an actual flood—which is precisely what the Genesis account provides. Genesis is written as history and corresponds to reality. No unpredictable gods clutter the text, and details that can be tested in an age of scientific enlightenment are found to be wholly plausible. For example, in contrast to the cube-shaped ark of Gilgamesh, modern engineering standards have proven the ark of Genesis to be ideally suited for floating and stability.

Finally, corrupted Flood accounts remind us that the reality of a great Flood is imprinted on the collective consciousness of virtually every major civilization from the Sumerian epoch to the present age. Although these corrupted accounts view the waters of the Flood through the opaque lens of paganism, they nevertheless lend credence to the occurrence of an actual flood.

In short, corrupted accounts such as the Epic of Gilgamesh provide independent confirmation of a vast flood in ancient Mesopotamia, complete with a Noah-like figure and an ark.

"I am going to bring floodwaters on the earth to destroy all life under the heavens, every creature that has the breath of life in it. Everything on earth will perish."

GENESIS 6:17

For further study, see Hank Hanegraaff, *Has God Spoken? Proof of the Bible's Divine Inspiration* (Nashville: Thomas Nelson, 2011).

Does Genesis Confirm the Reality of a Global Flood?

Even a cursory Google search makes plain that the Genesis account of the Flood is one of the favorite targets of those who do not believe the Bible. "If the Flood covered the mountains," they sneer, "it would put sea level at 29,055 feet, where everything on the ark would have frozen to death and not had enough oxygen to breathe."

First, the presumption that the floodwaters rose to 29,055 feet (twenty feet higher than the elevation of Mount Everest) is misguided at best. Given the science of plate tectonics, it is abundantly clear that Everest is significantly higher today than it would have been at the time of the great Flood.

Furthermore, the biblical text is not designed to communicate whether the Flood was global with respect to the earth or universal with respect to humanity. That debate is ultimately settled by a proper "reading" of the book of nature* (Psalm 19:1–4).

Finally, since civilization was largely confined to the Fertile Crescent*, we need not automatically presume that the floodwaters covered the globe. When Scripture tells us that "the whole world sought audience with Solomon to hear the wisdom God had put in his heart" (1 Kings 10:24), only the most ardent fundamentalist supposes this to include aborigines from Australia and indigenous peoples of the Americas. Of one thing we can be certain: The text of Scripture, both Old and New, communicates the reality of a great Flood in which "only a few people, eight in all, were saved" (1 Peter 3:20).

For further study, see Hank Hanegraaff, *Has God Spoken? Proof of the Bible's Divine Inspiration* (Nashville: Thomas Nelson, 2011).

Did Noah Take Seven Pairs or Just Two Pairs of Animals with Him on the Ark?

In his book *Jesus, Interrupted*, Bart Ehrman, distinguished professor of religious studies at the prestigious University of North Carolina, Chapel Hill, is perplexed about the number of animals Noah took with him on the ark. He poses the following question: "Does [Noah] take seven pairs of all the 'clean' animals, as Genesis 7:2 states, or just two pairs, as Genesis 7:9–10 indicates?"

First, I would like to pose a different question: Does it seem reasonable to suppose that an author capable of writing a masterpiece such as the book of Genesis would get confused within the span of several sentences, or is it more likely that Ehrman is straining at gnats and swallowing a camel?

Furthermore, is Ehrman's question legitimate, or has he created a problem out of whole cloth? Ehrman, of course, has created a fictional problem. Genesis 7:9–10 does not say that Noah was to take *just* two pairs.

Finally, if Ehrman really wants an answer, all he needs to do is consider the context. Several verses back, God said to Noah, "You are to bring into the ark two of all living creatures, male and female" (6:19). In Genesis 7:2–3, God added: "Take with you seven of every kind of clean animal, a male and its mate, and two of every kind of unclean animal, a male and its mate, and also seven of every kind of bird, male and female, *to keep their various kinds alive throughout the earth*." Together these verses provide a sufficient answer.

> Noah was six hundred years old when the floodwaters came on the earth. And Noah and his sons and his wife and his sons' wives entered the ark to escape the waters of the flood. Pairs of clean and unclean animals, of birds and of all creatures that move along the ground, male and female, came to Noah and entered the ark, as God had commanded Noah. And after the seven days the floodwaters came on the earth.
>
> GENESIS 7:6–10

Is It Silly to Believe in Noah's Flood?

From the perspective of a biblical worldview, the Flood is *the most* catastrophic event in the history of humanity. From the perspective of Internet spoofers, it is also *the most* comical. But is it really silly?

First, common sense demands that we allow for both natural and supernatural explanations to make sense of the universe in which we live. Realities such as the origin of life and the phenomenon of the human mind pose intractable difficulties for merely natural explanations.

Furthermore, reason forces us to look beyond the natural world to a supernatural Designer who not only sustains the world but supernaturally intervenes in the affairs of his created handiwork—which is precisely what the Genesis Flood account entails.

Finally, if we are willing to believe that God created the heavens and the earth—as opposed to the untenable notion that nothing created everything, that life came from nonlife, and that nonlife gave rise to objective morals—we will have little difficulty believing the Genesis account of Noah and the Flood.

For further study, see Hank Hanegraaff, *Has God Spoken? Proof of the Bible's Divine Inspiration* (Nashville: Thomas Nelson, 2011).

Creation and the Flood:
Advanced Reading

Is the Canopy Theory Credible?

The canopy theory posits that a blanket of water vapor separated the water under the sky "from the water above it" (Genesis 1:7) until the time of the Flood. This theory is used to account for everything from the water necessary for a global flood to the greenhouse effect necessary for human longevity. But does this theory correspond to reality?

First, the death knell to the canopy theory is that the "waters above the skies" and the expanse holding them back are found in the biblical text *after* Noah's Flood just as they were *before* it (e.g., Psalm 148:4–6).

Furthermore, to misinterpret Scripture is to miss its meaning. When Scripture asks, "Who can tip over the water jars of the heavens when the dust becomes hard and the clods of earth stick together?" (Job 38:37–38), we do violence to the text by positing an ill-conceived "canister" theory. A "canister" theory, like a canopy theory, simply does not correspond to reality.

Finally, as aptly noted in some young-earth

creationist literature, a blanket of water vapor prior to the Flood would have made the surface of the earth "intolerably hot, so a vapor canopy could not have been a significant source of the flood waters."

Praise him, you highest heavens
and you waters above the skies.
Let them praise the name of the LORD,
for he commanded and they were created.
He set them in place for ever and ever;
he gave a decree that will never pass away.

PSALM 148:4–6

Did Demons Have Sexual Relations with Women in Genesis 6?

Genesis 6:4 is one of the most controversial verses in the Bible. As with any difficult section of Scripture, it has been subject to a wide variety of interpretations. It is my conviction, however, that individuals who hold consistently to a biblical worldview must reject the notion that women and demons can engage in sexual relations. I reject this interjection of pagan superstition into the Scriptures for the following reasons.

First and foremost, the notion that demons can "produce" real bodies and have real sex with real women would invalidate Jesus' argument for the authenticity of his resurrection. Jesus assured his disciples that "a ghost does not have flesh and bones, as you see I have" (Luke 24:39). If indeed a demon could produce flesh and bones, Jesus' argument would be not only flawed but also misleading. In fact, it might be logically argued by extension that the disciples did not see the post-resurrection appearances of Christ but rather a demon masquerading as the resurrected Christ.

Furthermore, demons are nonsexual, nonphysical beings and, as such, are incapable of having sexual relations and producing physical offspring. To say that demons can create bodies with DNA and fertile sperm is to say that demons have creative power—which is an exclusively divine prerogative. Moreover, if demons could have sex with women in ancient times, we would have no assurance they could not do so in modern times. Nor would we have any guarantee that the people we encounter every day are fully human. While a biblical worldview does allow for fallen angels to *possess* unsaved human beings, it does not support the notion that a demon-possessed person can produce offspring that are part-demon, part-human. In fact, Genesis 1 makes it clear that all God's living creations are designed to reproduce "according to their kinds" (verses 21, 24, 25).

Finally, this mutant theory prompts serious questions about the spiritual accountability of hypothetical demon-humans and their relation to humanity's redemption. Angels rebelled individually, were judged individually, and were offered no plan of redemption in Scripture. On the other hand, humans fell corporately in Adam, are judged corporately in Adam, and are redeemed corporately through Jesus Christ. We have no biblical way of determining what category the demon-humans would fit into—whether they would be judged as angels or as men and, more significantly, whether or not they are even among those for whom Christ died.

The better interpretation is that "sons of God" refers to the godly descendants of Seth and "daughters of men" to the ungodly descendants of Cain. Their cohabitation caused humanity to fall into such utter depravity that God said, "I will wipe mankind, whom I have created, from the face of the earth" (Genesis 6:7).

Creation
and the Age
Question

Is This a Young World After All?

Many Christians consider the universe to be relatively young. But is it really? While special revelation (the Bible) does not specifically address the age issue, general revelation (the book of nature*) provides a veritable host of clues.

First, the reality that nothing travels faster than the *speed of light* and that billions of light-years separate us from distant galaxies leads logically to the assumption that the age of the universe is measured in billions rather than in thousands of years.

Furthermore, *star life* is a persuasive argument for a universe measured in billions of years. Star life depends on star mass. A star like the sun has enough fuel to burn for an estimated 9 billion years. Conversely, the fuel of a star half the size of the sun may last as long as 20 billion years. As such, the universe is presumed to be at least as old as the oldest stars within in it. While biological history may be a matter of inference, astronomical history is a function of direct observation. Put another way, star formation can be observed in all of its stages.

Finally, *sequential layers* in the formation of ice cores in places such as Antarctica and Greenland point to an earth much older than six thousand years. Just as arborists count tree rings to estimate the age of trees, researchers count sequential layers to date the age of ice cores. This data log seems to demonstrate that the age of the earth is at least hundreds of times older than the age presumed by young-earth creationists[*].

For further study, see Guillermo Gonzalez and Jay W. Richards, *The Priveleged Planet: How Our Place in the Cosmos Is Designed for Discovery* (Washington, DC: Regnery, 2004).

When Was the
Universe Created?

Using the speed of light (186,000 miles per second) in their calculations, astronomers have determined that the observable universe with its 100 billion galaxies, each containing a 100 billion stars, is at least 15 billion light-years in diameter. (A light-year is the distance light travels in a year—a distance measured not in billions but trillions of miles—about 5,878,499,810,000 miles!)

Furthermore, the age of the universe is measured in billions of years due to what is popularly referred to as the redshift of the galaxies: the reddish light marks motion away from the earth much like the audible pitch of a train shifts as it moves off into the distance. The redshift that marks galaxies as they move apart at the speed of light allows astronomers to extrapolate backward billions of years to the point at which the "stretching of space" began.

Finally, science points to realities such as background radiation, radioactive decay, entropy*, star ages, and white dwarf stars as proof positive that the universe is billions of years old. (For example, a star becomes a white dwarf—essentially a dead star—only after billions of years of nuclear fusion and subsequent cooling.) These multiple sets of independent empirical* evidence all converge on a limited range of dates for the origin of the universe, a date somewhere between 10 and 20 billion years ago.

It is important to note that even 10–20 billion years is insufficient for the evolution of a protein molecule, much less a living cell.

Were the Genesis Creation Days Literal, Long, or Literary?

There are three dominant schools of thought within evangelical Christianity regarding the Genesis days of creation.

First, the popular *twenty-four-hour* view posits that God created the heavens and the earth in six sequential, literal days. A majority in this camp view the universe to be approximately six thousand years old and consider all death, including animal death, to be a direct function of Adam's fall.

Furthermore, the *day-age* perspective suggests that God created the heavens and the earth in six long, sequential "days"—with each "day" totaling billions of years. In contrast to the twenty-four-hour perspective, the day-age view maintains that human and animal suffering and death are the result of God's "very good" creation prior to Adam's fall into a life of perpetual sin terminated by death.

Finally, the *framework* perspective holds that the seven days of creation are nonliteral and nonsequential but nonetheless historical. In concert with the day-age perspective, the framework perspective views animal death before the fall as consistent with the goodness of God's creation, and believes that the age question is settled by natural revelation (the book of nature*) rather than by special revelation (the Bible).

In my view, the literary-framework interpretation most closely corresponds to reality—though I cannot abide animal death prior to the fall as consistent with a "very good" creation (see "Could Carnivores and Catastrophes Have Existed Prior to the Fall?" on page 27).

> For in six days the Lᴏʀᴅ made the heavens and
> the earth, the sea, and all that is in them, but he
> rested on the seventh day.
>
> Exᴏᴅᴜs 20:11

For further study, see *The Genesis Debate: Three Views on the Days of Creation*, ed. David G. Hagopian (Mission Viejo, CA: Crux Press, 2001).

Did God Create His Handiwork with the Appearance of Age?

It is frequently argued that God created the universe and all it entails with the appearance of age. Does this notion correspond to the reality of Scripture and science?

First, we should note that the Bible doesn't answer the age question. Some creationists suggest that God created Adam with the appearance of age. In reality, we simply do not know. Was Adam created with calluses on his feet? Did he have a belly button? Was he fashioned replete with childhood memories? One would think not, but the Bible simply doesn't say.

Furthermore, the notion that God created his handiwork with the appearance of age is logically unfalsifiable. In other words, you can neither prove it nor disprove it. For example, how could you prove false the notion that you were created five seconds ago and that your recollection of the previous paragraph is just an implanted memory?

Finally, consider an observable astronomical event such as *Supernova 1987A*—an event with an identifiable "before" *and* "after." Prior to 1987, this supernova was a star in a distant galaxy 168,000 light-years away. On February 23, 1987, however, the star exploded and became a supernova. In other words, 168,000 years ago the star exploded, and in 1987 the light of that event finally reached earth—unless, of course, God created the universe six thousand years ago. Then the supernova might well be likened to a documentary film of an event that never really happened.

In sum, the notion that the universe is not authentically old but merely appears to be old creates more conundrums than it solves. Indeed, what good teacher would ask students to put faith in a textbook intentionally filled with lies?

Creation and the Age Question: Advanced Reading

What Are Exegetical Liabilities of the Twenty-Four-Hour View?

Just as there are extrabiblical liabilities for the twenty-four-hour view (speed of light, star life, etc.), so too there are exegetical liabilities*.

First, young-earth creationists* suggest that God employed nonsolar light to govern the days until he created the sun, moon, and stars on Day 4. Yet the biblical text literally says, "There was evening, and there was morning," indicating that the first three days of creation were normal solar days encompassing daylight and darkness (Genesis 1:5, 8, 13).

Furthermore, the dominant argument that the Hebrew word *yom* (meaning *day*) used with a numeral always, always, always refers to a literal twenty-four-hour solar day does not correspond to reality. Hosea 6:2 is a devastating counter-example. Here, as in other passages (Zechariah 14:7), *yom* preceded by a numeral represents a period of time far longer than a single solar day.

Finally, the unending nature of the seventh day constitutes a major exegetical problem for the twenty-four-hour interpretation. Logically and literarily, the seventh day cannot simultaneously be unending *and* temporal.

In sum, the days of Genesis are rendered literal solar days not to establish a chronology of creation, but to remember the purposes of God in creation.

For further study, see *The Genesis Debate: Three Views on the Days of Creation*, ed. David G. Hagopian (Mission Viejo, CA: Crux Press, 2001).

Are There Gaps in the Genealogies of Genesis?

According to the famed chronology of Bishop James Ussher, the creation of Adam can be dated to precisely 4004 BC based on Old Testament genealogies such as those provided by Moses in Genesis 5 and 11. Thus, we can be certain that the universe is six thousand years old. Or can we?

First, in comparing Scripture with Scripture, we immediately recognize an omission in the genealogy of Genesis 11. In the genealogy found in Luke 3, *Cainan* is listed between *Shelah* and *Arphaxad.* In the Genesis 11 genealogy, Cainan is missing. While some have ascribed this to a copyist error, all extant biblical manuscripts containing the genealogy—except two ancient witnesses—include the name. As such, through the tenacity of the text, the autograph has emerged replete with Cainan as a legitimate inclusion in the genealogy.

Furthermore, the genealogies of Genesis, in concert with other biblical genealogies, are symmetrical and deliberately arranged. Matthew, for example, skillfully organized the genealogy of Jesus into three groups of fourteen, the numerical equivalent of the Hebrew letters in King David's name ($4+6+4=D+V+D$). Thus, Matthew's genealogy simultaneously highlights the most significant names in the lineage of Jesus and artistically emphasizes our Lord's identity as Messiah, who forever sits upon the throne of David. Like Matthew, Moses skillfully organized the royal genealogy from which Jesus emerged into two symmetrical groups—ten generations before the Flood (Genesis 5) and ten after the Flood (Genesis 11). Thus, there is ample precedent for seeing the genealogies as symmetrical rather than sequential.

Finally, when Matthew 1:8 lists Jehoram as the father of Uzziah, three generations are bypassed (Ahaziah, Joash, and Amaziah). Thus, far from being the father of Uzziah, Jehoram fathered the *line* that culminated in Uzziah. This sort of telescoping has significant precedent in biblical history. In the book of Daniel, for instance, Belshazzar is called the son of Nebuchadnezzar (5:2) when in reality he was the son of Nabonidus, the son-in-law of Nebuchadnezzar.

There is a good reason biblical scholars are obsessed with genealogies. While they do not provide a precise chronology between the first and the last Adams, they are rife with theological significance.

What About Progressive Creationism?

Progressive creationists*, like their young-earth counterparts, face significant challenges from both science and Scripture.

First, the quest to find maximal harmony between science and Scripture has led to forced interpretations of both. On the one hand, Scripture is harmonized with science through the contention that the sun and stars were created prior to Day 1—but were not visible from the surface of the earth until Day 4. On the other hand, science is harmonized with Scripture through a strained explanation that allows vegetation to thrive during a geological age in which there is no direct sunlight.

Furthermore, in order to harmonize Scripture with modern science, progressive creationists* compromise the moral significance of natural evil. As such, they ascribe all morally meaningful evil to that which takes place after the initial human sin. Though progressive creationists have acknowledged that animals "manifest attributes of mind, will, and emotions" and are "uniquely endowed with the capacity to form relationships—with each other and with humans," the horrors of suffering, death, and extinction are said to be part and parcel of God's "very good" creation.

Finally, from a progressive creationist perspective, even Eden was not immune to the havoc of parasitism, animal deaths, infections, and natural catastrophes. Eve herself suffered the ravages of pain, which only "increased" as a result of her sin. To put it bluntly, there never was a perfect paradise. Worse yet, God is implicated in using disease, decay, destruction, and even death (dubbed "random, wasteful inefficiencies") as the means by which to bring about the "very good" world in which Eve experienced pain. As Dr. William Dembski aptly noted, "The difficulty with this suggestion, which is made throughout the old-earth creationist literature, is that natural evil becomes simply a tool for furthering God's end rather than a consequence of human sin. Old-earth creationism thus opens God to the charge of inflicting pain simply to advance a divine agenda."

In reality, there is simply no need for far-fetched special pleading. A proper reading of the book of nature* reveals natural evil before Adam.

Likewise, a proper reading of sacred Scripture reveals an infinite God who redeems humankind by acting across time. As the effects of the cross are retroactive, so too the effects of the fall are retroactive.

> . . . the Lamb that was slain from the creation of the world.
>
> REVELATION 13:8

For further study, see William A. Dembski, *The End of Christianity: Finding a Good God in an Evil World* (Nashville: B&H Publishing, 2009).

What About Trying to Find Contemporary Scientific Paradigms in Scriptural Passages?

Attempts to read scientific paradigms into scriptural passages have given Christianity one black eye after another. More often than not, a serious misunderstanding of the art and science of biblical interpretation is at the root of the problem.

First, churchmen once taught that the earth was stationary on the basis of Psalm 93:1—"The world is firmly established; it cannot be moved." Clearly, this is not the intent of the passage. A quick look at the context reveals the meaning: the kingdom of the "Lord [who] is robed in majesty" (hardly a comment on his clothing) cannot be shaken by the pseudo-powers of earth.

Furthermore, in an attempt to find concord between science and Scripture, Isaiah likewise has been robbed of meaning and magnificence. Young-earth creationists* argue that Isaiah 40:22

teaches sphericity ("circle of the earth"); progressive creationists allege it points to Big Bang cosmology ("stretches out the heavens"); and anti-creationists assert it supports flat earth mythology ("spreads [the heavens] out like a tent"). One can only imagine theistic evolutionists averring that humans evolved from insects ("people are like grasshoppers").

Finally, with astronomy texts in hand, leaders in the progressive creationist* movement contend that the book of Job, arguably the oldest in the Bible, contains a stunning reference to the Big Bang: "He alone stretches out the heavens and treads on the waves of the sea" (9:8). In like fashion, skeptics suppose an un-evolved Job thought earth was set on pillars: "He shakes the earth from its place and makes its pillars tremble" (9:6).

While the earth is spherical and Big Bang cosmology does accord well with the opening of Genesis, neither Isaiah's reference to "the circle of the earth" nor Job's comment about stretching is a basis for harmonizing modern cosmology with the Genesis account of creation.

Can Radiometric Dating Be Trusted?

A major contention of young-earth creation-ists* is that radiometric dating* (measuring radioactive decay) is not reliable because the rate of nuclear decay was greater in the past than it is in the present.

First, this contention creates more problems than it solves. Simply put, the nuclear radiation rate necessary to make sense of a six-thousand-year-old universe would have been lethal to plant, animal, and human life.

Furthermore, physicists have not simply pre-sumed decay rates to be consistent. They have made a concerted effort to disprove radiometric dating by subjecting radioactive atoms to extreme temperatures, extreme pressures, and a variety of electromagnetic variations. To date, however, no change in the rate of decay of any geologically sig-nificant radioactive isotope has been discovered.

Finally, the age of the earth as determined through radiometric dating processes corresponds to age parameters projected by such astronomical measurements as star life, which also demonstrates the age of the earth to be hundreds of times older than that presumed by young-earth creationists.

One thing is certain. Present projections regarding the age of the earth are wholly insufficient for the evolution of even the most basic of all protein molecules. Indeed, present age projections wholly undermine the evolutionary hypothesis.

> He stood, and shook the earth;
> he looked, and made the nations tremble.
> The ancient mountains crumbled
> and the age-old hills collapsed.
> His ways are eternal.
>
> HABAKKUK 3:6

For further study, see Davis A. Young and Ralph F. Stearly, *The Bible, Rocks, and Time: Geological Evidence for the Age of the Earth* (Downers Grove, IL: IVP Academic, 2008).

Creation and
the Problem
of Evil

Is Animal Suffering a Consequence of Adam's Sin?

"I cannot persuade myself," wrote Darwin, "that a beneficent and omnipotent God would have designedly created [the *Ichneumonidae*] parasitic wasps with the express intention of their feeding within the living bodies of caterpillars." This conundrum ultimately led Darwin to dispense with the notion of a Creator God. In reality, however, Adam—not the Almighty—bears responsibility for the origin of moral and natural evil in the world.

First, the Bible clearly teaches that "sin entered the world through one man, and death through sin" (Romans 5:12). As a result, the whole of creation was subjected to "frustration" and "decay" (see Romans 8:19–23; cf. Genesis 1:29–30; 9:1–4; Psalm 104:19–28).

Furthermore, the federal headship* of Adam (Romans 5:12–21; 1 Corinthians 15:20–26) extends beyond humanity to *all* of God's creation. Even the ground was cursed as a direct result of Adam's rebellion. Not only so, but the present curse and the promised redemption extend beyond the ground to the very animals that walk upon it (Isaiah 11:6–9; 65:25; Revelation 21–22).

Finally, far from dispensing with God as a result of contemplating such natural horrors as a parasitic wasp, human and animal suffering should have driven Darwin to contemplate the full consequences of alienation from God. Indeed, exposure to natural evil outside the comforts of the garden must surely have caused Adam to understand the full gravity of his fall from grace. Put another way, chaos outside the garden reflected the horror of Adam's sin-sick soul.

Tragically, Darwin could only conceive of time as linear. Had he comprehended a God unbounded by time, his evolutionary hypothesis may never have taken root. Surely God could cause the effects of the fall to temporally precede their cause! As intelligent design* theorist Dr. William Dembski has well said, "Just as the death and resurrection of Christ is responsible for the salvation of repentant people throughout all time, so the fall of humanity in the garden of Eden is responsible for every natural evil throughout all time (future, present, past, and distant past preceding the fall)."

The creation waits in eager expectation for the sons of God to be revealed. For the creation was subjected to frustration, not by its own choice, but by the will of the one who subjected it, in hope that the creation itself will be liberated from its bondage to decay and brought into the glorious freedom of the children of God.

We know that the whole creation has been groaning as in the pains of childbirth right up to the present time. Not only so, but we ourselves, who have the firstfruits of the Spirit, groan inwardly as we wait eagerly for our adoption as sons, the redemption of our bodies.

ROMANS 8:19–23

For further study, see William A. Dembski, *The End of Christianity: Finding a Good God in an Evil World* (Nashville: B&H Publishing, 2009).

Could Carnivores and Catastrophes Have Existed Prior to the Fall*?

The sharpest point of contention between young- and old-earth creationists is death before Adam. Young-earth creationists* maintain that all death—including animal death—is a function of the fall. Old-earth creationists* believe that carnivorous creatures and natural catastrophes existed prior to the sin of Adam as part and parcel of God's "very good" creation. Perhaps this rift is unnecessary.

First, the false assumption underlying this debate is that Adam's sin necessarily precedes the advent of all evil (except for the rebellion of Satan). Thus, young-earthers are constrained to reject standard astronomical and geological dating, while old-earthers are compelled to reinterpret biblical passages that speak of natural evil as a direct consequence of human sin (Genesis 3:14–19; Romans 5:12–21; 8:18–25; 1 Corinthians 15:20–23). In reality, orthodoxy demands faithfulness to truth in both general and special revelation.

Furthermore, an orthodox understanding of the Creator of all things is that he acts *transtemporally* ("across time") in all of his creation. Thus, the cross of Christ retroactively atones for Adam's sin, just as it proactively atones for the sin of Adam's yet unborn descendants. In similar fashion, there is little difficulty conceiving of a transcendent God who predestines natural evil to precede the fall even though the fall is the necessary cause of the evils that precede it.

Finally, while it is natural for humans to see time's arrow from a forward perspective, we do well to recognize that God's infallible truth is not always presented in chronological fashion. As such, the advent of the last Adam not only atones for sin thousands of years after the advent of the first Adam, but the last Adam (Jesus) is biblically portrayed as the Lamb "slain from the creation of the world" (Revelation 13:8).

In sum, God's purposes are presented in both chronological as well as kairological* fashion. In other words, we see events reported in the order in which they happened, and we see events recorded according to their purpose and significance. "Before they call I will answer," says the Almighty, "while they are still speaking I will hear" (Isaiah 65:24).

> "I am God, and there is no other;
> I am God, and there is none like me.
> I make known the end from the beginning,
> from ancient times, what is still to come.
> I say: My purpose will stand,
> and I will do all that I please."
>
> ISAIAH 46:9–10

For further study, see William A. Dembski, *The End of Christianity: Finding a Good God in an Evil World* (Nashville: B&H Publishing, 2009).

How Could a Good God Create a World in Which Things Go Desperately Wrong?

At first blush, it may seem as though there are as many responses to this question as there are religions. In reality, there are only three. Pantheism denies the existence of good and evil because god is all and all is god. Philosophical naturalism* (the worldview undergirding evolutionism) supposes that everything is a function of natural processes, such as brain chemistry and genetics, thus there is no good and evil. Theism alone has a relevant response—and only *Christian* theism offers one that is satisfactory.

First, Christian theism acknowledges that God created the *potential* for evil because God created humans with freedom of choice. We choose to love or to hate, to do good or to do evil. The record of history bears eloquent and chilling testimony to the fact that we humans have, of our own free will, actualized the reality of evil through our choices.

Furthermore, without choice, love is rendered meaningless. God neither forces his love on people nor forces people to love him. Instead, God, the supreme exemplar of love, created us with freedom of choice. Without such freedom, we would be little more than preprogrammed robots.

Finally, the fact that God created the potential for evil by granting us freedom of choice will ultimately lead to the best of all possible worlds—a world in which "there will be no more death or mourning or crying or pain" (Revelation 21:4). Those who choose Christ will be redeemed from evil by his goodness and will forever be able to *not* sin.

We know that in all things God works for the good of those who love him, who have been called according to his purpose.

ROMANS 8:28

For further study, see Joni Eareckson Tada and Steven Estes, *When God Weeps: Why Our Suffering Matters to the Almighty* (Grand Rapids: Zondervan, 1997).

Creation and the Problem of Evil: Advanced Reading

How Do Progressive Creationists Deal with the Problem of Natural Evil Prior to Adam?

Progressive creationists* believe that animal suffering and death existed for millions of years prior to the creation of Adam and Eve. Thus the question: isn't natural evil, like moral evil, a consequence of human sin?

First, old-earth progressive creationists see natural catastrophes and carnivorous animals as part of God's "very good" creation. As such, they render animal suffering and death as good rather than evil.

Furthermore, progressive creationists make a distinction between the words *good* and *perfect*. Thus, in their view, God created the heavens and the earth and all that is in them "very good"—not perfect.

Finally, progressive creationists deny a biblical basis for believing that nature red in tooth and claw* resulted from Adam's fall. Thus, they relate passages such as Romans 5:12 and 1 Corinthians 15:21–22 solely to human suffering and death, not the death and suffering evident in nature.

> Sin entered the world through one man, and death through sin, and in this way death came to all men.
>
> ROMANS 5:12

For further study, see "Could Carnivores and Catastrophes Have Existed Prior to the Fall?" on page 127.

Creation and
Dinosaurs

Is There Evidence That Humans and Dinosaurs Walked Together?

Maybe you've seen it on the Internet, the popular claim that human footprints have been discovered alongside dinosaur tracks in the Paluxy Riverbed of Glen Rose, Texas. It is argued that these footprints in stone prove that the world is six thousand years old and therefore dinosaurs existed contemporaneously with human beings. Intensive scrutiny, however, has caused many to abandon this pretext.

First, while footprints of three-toed carnivorous dinosaurs (tridactyls) are evidenced in stone, there is no compelling evidence for human footprints in the same time and space. Indeed, the prints are either too large to be human or questionable in origin.

Furthermore, some of the prints previously attributed to humans show claw marks consistent with three-toed tridactyls and inconsistent with five-toed humans.

Finally, many prints claimed to be human are little more than erosion patterns—an illustration of wish giving birth to reality.

Creationists who have questioned the Paluxy Riverbed prints are to be commended for following facts and tracks wherever they lead. As noted by the Institute for Creation Research in their *Acts and Facts*, "Scientists must always be willing to re-evaluate prior interpretations once new data becomes available. Creationists have rightly accused evolutionists of being close-minded on key issues, and we cannot afford to become like them in this respect. Jesus Christ claimed to be the Truth, and since we follow Him, we must be lovers of truth."

Do Recently Discovered T. Rex Bones Point to a Young Earth?

The Internet is awash with evidence that scientists have found "unfossilized" T. rex bones replete with "*fresh* blood cells and hemoglobin." Thus, we now have "powerful testimony against the whole idea of dinosaurs living millions of years ago." Or do we?

First, the scientists who conducted the research never claimed to have found either fresh blood cells or hemoglobin. The scientific journals instead report the discovery of collagen, a protein found in bone. And the presence of collagen would not, in and of itself, prove that humans walked with dinosaurs in the last several thousand years.

Furthermore, headlines such as "Blood Chemicals Found in Dino Bone" are grounded in sensationalistic news flashes, not scientific literature. And yet this "fact" of science is cited as one of the six evidences that the universe is young.

Finally, statements that no intact dinosaur blood cells have been discovered have prompted the unfounded allegation that scientists are presently engaged in a massive cover-up or conspiracy. This, however, is no more factual than the contention that we now have "physical evidence that dinosaur bones are not millions of years old."

There is no need or room for such make-believe stories in the creation/evolution debate. In truth, an earth measured in billions of years is no threat to a biblical view of creation. No matter how many years the evolutionist postulates, chance operating on natural processes could not produce so much as a DNA molecule, let alone a dinosaur.

Test everything. Hold on to the good.

1 THESSALONIANS 5:21

Creation and Dinosaurs:
Advanced Reading

Are Behemoth and Leviathan Dinosaurs?

Behemoth (Job 40) and Leviathan (Job 41) are frequently referenced as evidence that the patriarch Job lived alongside such dinosaurs as brachiosaurus and kronosaurus. Is this fact or fiction?

First, it is important to recognize that Job serves as a literary polemic*, a written indictment, against the gods of ancient Near Eastern mythology. Not only were pagans *literally* supplanted by people of the promise, but their mythological narratives were *literarily* supplanted by the meta-narrative*, the overarching story of Scripture. It is not Baal who destroys "the writhing serpent, encircler-with-seven-heads" (Ugaritic text) but Yahweh who "crushed the heads of Leviathan" (Psalm 74:14). Thus, through literary subversion*, God recasts pagan myth in a manner that corresponds to reality.

Furthermore, it is crucial to note the literary progression of Job. After thirty-plus chapters of rambling human speculations, God answers Job out of the storm. In essence, the Almighty asks Job if he would like to try his hand at running the universe for a while: "Who fathers the drops of dew?" "Who can tip over the water jars of the heavens?" "Do you give the horse his strength?" "Does the eagle soar at your command?" Consider Behemoth, who "ranks first among the works of God" or the sea dragon—"Can you pull in the leviathan with a fishhook?" The literary progression moves from creation, to creatures, to the cherub who once ranked first in the order of creation. To Job, the primal monster of the land, like the primal monster of the sea, was indomitable. To Jehovah, Behemoth and Leviathan were mere pets on a divine leash. In the literary progression of the Bible, the monster is vanquished. Says Isaiah, "Leviathan the gliding serpent, Leviathan the coiling serpent; [God] will slay the monster of the sea" (27:1).

Finally, in interpreting Scripture in light of Scripture, the literary personification of Satan becomes readily apparent. In Genesis he is presented as an alluring serpent that tempts humanity to fall into lives of perpetual sin terminated by death; in Psalms he is portrayed as a multiheaded monster opposing the purposes of God; in Isaiah he is a coiling serpent emerging out of the primal waters; and in Revelation, a red dragon that personifies the extremities of evil.

In sum, Leviathan and Behemoth are not dinosaurs but personifications that illustrate a metaphysical reality. As such, the mythology of the dragon underscores the reality of the devil.

> I saw an angel coming down out of heaven, having the key to the Abyss and holding in his hand a great chain. He seized the dragon, that ancient serpent, who is the devil, or Satan, and bound him for a thousand years.

REVELATION 20:1–2

Is There a Biblical Distinction Between Dinosaurs and Dragons?

Young-earth creationists* often argue that the Bible is full of references to dinosaurs. Nineteenth-century English paleontologist Sir Richard Owen did not coin the word *dinosaur* ("terrible lizard") until 1841, thus the King James Version erroneously substituted the word *dragon* for *dinosaur*. Had the word been coined prior to 1611, passages such as Isaiah 27:1 and Psalm 74:13 would reference dinosaurs instead of dragons. Or would they?

First, a quick look at passages cited by young-earth creationists is sufficient to recognize that they have been misused as pretexts for the notion that dinosaurs lived contemporaneously with humans. Isaiah 27 depicts the destruction of demonic nations that presume to frustrate the purposes of God. As such, God will destroy "Leviathan the gliding serpent, Leviathan the coiling serpent; he will slay the monster of the sea" (v. 1). Likewise, in

Psalm 74, God crushes Leviathan—multiheaded monster of the sea—and promises the establishment of a new world order in which "there will be no more death or mourning or crying or pain" (Revelation 21:4). To recast multiheaded demonic forces in the image of single-headed dinosaur fossils is singularly wrongheaded.

Furthermore, it should be noted that the motif of a sovereign God who rides the stormy clouds and destroys the multiheaded monster arising out of the chaotic seas is a shared ancient Near Eastern narrative appropriated by biblical writers as a means of arguing against the false gods of paganism. Such literary substitution is employed by the inspired biblical authors to underscore the reality that it is not Baal who rules supreme from Mount Saphon (in western Syria) but Yahweh the king of Mount Sinai who conquers the roiling habitat of the coiling serpent.

Finally, just as biblical evidence demonstrating that dinosaurs lived in the same time and space as humans is suspect, so too the extrabiblical evidence. As demonstrated elsewhere, footprints of humans together with dinosaurs in the Paluxy Riverbed of Glen Rose, Texas, have proven suspect, as have alleged *fresh* blood cells in the unfossilized remains of a T. rex (see "Do Recently Discovered T. Rex Bones Point to a Young Earth?" on page 140).

In sum, an objective reading of the biblical text is sufficient to demonstrate that dragons are not equivalent to dinosaurs. Abusing biblical truth is no small matter.

For further study, see "Is there Evidence That Humans and Dinosaurs Walked Together?" (p. 138); also "Do Recently Discovered T. Rex Bones Point to a Young Earth?" (p. 140).

Is *Archaeopteryx* the Missing Link Between Dinosaurs and Birds?

Whenever I say there are no fossil transitions[*] from one species to another, someone inevitably brings up *Archaeopteryx*. This happens so frequently that I have decided to coin a word for the experience: *pseudosaur*. *Pseudo* means false and *saur* refers to a dinosaur or a reptile (literally, lizard). Thus, a *pseudosaur* is a false link between reptiles (such as dinosaurs) and birds. Myriad evidences demonstrate conclusively that *Archaeopteryx* is a full-fledged bird, not a missing link between birds and dinosaurs.

First, fossils of both *Archaeopteryx* and the kinds of dinosaurs *Archaeopteryx* supposedly descended from have been found in a fine-grained German limestone formation said to be Late Jurassic. (The Jurassic period is said to have begun approximately 200 million years ago, lasting some 50 million years.) Thus, *Archaeopteryx* is not a likely candidate for the missing link, since birds and their alleged ancestral dinosaurs thrived during the same period.

Furthermore, initial *Archaeopteryx* fossil finds gave no evidence of a bony sternum, which led paleontologists to conclude that *Archaeopteryx* could not fly or was a poor flier. However, in April 1993, a seventh specimen was reported that included a bony sternum. Thus, there is no further doubt that *Archaeopteryx* was as suited for power flying as any modern bird.

Finally, to say that *Archaeopteryx* is a missing link between reptiles and birds, one must believe that scales evolved into feathers for flight. Air friction acting on genetic mutation supposedly frayed the outer edges of reptilian scales. Thus, over the course of millions of years, scales became increasingly like feathers until one day the perfect feather emerged. To say the least, this idea must stretch the credulity of even the most ardent evolutionists.

These and myriad other factors overwhelmingly exclude *Archaeopteryx* as a missing link between birds and dinosaurs. The sober fact is that *Archaeopteryx* appears abruptly in the fossil record, with masterfully engineered wings and feathers common in the birds observable today. The late Stephen Jay Gould of Harvard and Niles Eldredge of the American Museum of Natural History, both militant evolutionists, have concluded that *Archaeopteryx* cannot be viewed as a transitional form.

> God made the wild animals according to their kinds,
> the livestock according to their kinds, and all the
> creatures that move along the ground according to their
> kinds. And God saw that it was good.
>
> GENESIS 1:25

For further study, see Hank Hanegraaff, *Fatal Flaws: What Evolutionists Don't Want You to Know* (Nashville: Thomas Nelson, 2008) and Jonathan Wells, *Icons of Evolution: Science or Myth? Why Much of What We Teach About Evolution Is Wrong* (Washington, DC: Regnery, 2000).

Is *Pro-Avis* a Missing Link Between Lizards and Birds?

A few years after Harvard's Stephen Jay Gould ruled out *Archaeopteryx* as a missing link, Yale's John Ostrom proposed *Pro-avis*. Is Pro-avis a function of science or just science fiction?

First, unlike *Archaeopteryx*, no fossil evidence exists for Pro-avis. Eminent paleontologist Dr. John Ostrom, credited for revolutionizing modern understanding of dinosaurs, simply pictured a Pro-avis prototype in *American Scientist,* and Pro-avis sprang into existence. With due credit to creativity, this is hardly credible.

Furthermore, to hold Pro-avis as the missing link between reptiles and birds requires believing that air friction acting on genetic mutation frayed the outer edges of reptilian scales. Thus, over the course of millions of years, scales were transformed into fantastic flying feathers.

Finally, as science advances, even the most doctrinaire evolutionists are coming to the realization that *pseudosaurs* (false lizards) like Pro-avis simply don't fly in an age of scientific enlightenment.

In fairy tales, frayed scales turn into feathers and frogs into princes. In macroevolution* all one needs to do is add millions of years and little Proavises turn into fantastic flying fowl.

Newsweek aptly summarized the sentiment of leading evolutionists at a conference in Chicago: "Evidence from fossils now points overwhelmingly away from the classical Darwinism which most Americans learned in high school."

For further study, see Hank Hanegraaff, *The FACE That Demonstrates the Farce of Evolution* (Nashville: Thomas Nelson, 2001).

Is It Reasonable to Believe That Scales Evolved into Feathers?

A primary evolutionary contention is that, in the course of merely millions of years, reptilian scales became more and more like feathers until, one day, the perfect feather emerged. Is this reasonable or ridiculous?

First, as science advances, it has brought to light an unanticipated world of enormous complexity that requires the evolutionist to take a huge leap of faith. The science of statistical probability alone demonstrates that chance operating in concert with undirected processes (given even millions of years) can no more create a finch's feather than it could a fish's fin.

Furthermore, the meticulous engineering of feathers hardly squares with evolutionary probabilities. The central shaft of a feather has a series of barbs projecting from each side at right angles. Rows of smaller barbules in turn protrude from both sides of the barbs. Tiny hooks (barbicels) project downward from one side of the barbules and interlock with ridges on the opposite side of adjacent barbules. There may be as many as a million barbules cooperating to bind the barbs into a complete feather, impervious to air penetration.

Finally, consider the profound aerodynamic properties of a feathered airfoil. The positioning of feathers is controlled by a complex network of tendons that allows the feathers to open like the slats of a blind when the wing is raised. As a result, wind resistance is greatly reduced on the upstroke. On the downstroke, the feathers close, providing resistance for efficient flight.

To attribute the fearsome flight of the falcon and the delicate darting flitter of the hummingbird to an unguided purposeless process is to fly from knowledge to anti-knowledge.

For further study, see Hank Hanegraaff, *The FACE That Demonstrates the Farce of Evolution* (Nashville: Thomas Nelson, 2001).

Creation and
Evolution

What Is Darwin's Tree of Life?

According to Charles Darwin's Tree of Life*, humans and fruit flies share a common ancestor. According to Richard Dawkins, so do boys and bananas. This begs the question: What is Darwin's version of the Tree of Life?

First, Darwin's Tree of Life is an illustration. It appears in *The Origin of Species* to persuade the faithful that all species are, as Darwin put it, "lineal descendants of some few beings which lived long before the first bed of the Cambrian system was deposited." At the root of the tree is a handful of organic building blocks; at the tips of its budding branches are all modern species.

Furthermore, the Tree of Life is an icon. And not just *an* icon—*the* principal symbol of evolution. Indeed, for multitudes this icon has become the argument. The mere mention of it invokes devotees to bow deeply at the twin altars of common descent* and natural selection*. "I should infer," the chief priest intoned, "that probably all the organic beings which have ever lived on this earth have descended from *some one* primordial form."

Finally, the Tree is incorrect. In the geological period designated Cambrian, the highest orders in the biological hierarchy appear suddenly and fully formed. As Oxford zoologist Richard Dawkins acknowledged, "It is as though [fossils] were just planted there, without any evolutionary history."

And that is precisely the case. Darwin's Tree of Life is not only uprooted by the Cambrian Explosion*, but the fossil record in general shows no evidence of the origin of species by means of common descent and natural selection.

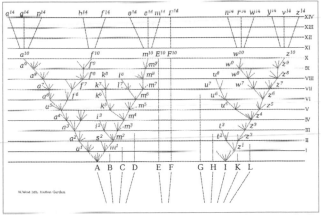

Darwin's Tree of Life

For further study, see Jonathan Wells, *Icons of Evolution: Science or Myth? Why Much of What We Teach About Evolution Is Wrong* (Washington, DC: Regnery, 2000).

Did Hippos Evolve into Whales?

In *The Origin of Species*, Darwin speculated that bears might well evolve into whales: "I see no difficulty in a race of bears being rendered by Natural Selection*, more and more aquatic in their structure and habits, with larger and larger mouths, till a creature was produced as monstrous as a whale." In stark contrast, contemporary Darwinists hold that hippo-like animals evolved into whales. Is this a hip potentiality, or a whale of a tale?

First, to believe that hippo-like animals evolved into whales takes an enormous leap of faith. Simply put, extant fossil transitions* are scant. As such, belief in the evolutionary development of physiological wonders such as blowholes, sonar, and diving mechanisms takes a commendable amount of Darwinian devotion.

Furthermore, molecular evidence appears to be at odds with fossil evidence. Fossil similarities point to hippos as first cousins of pigs, while molecular similarities position hippos as first cousins to whales. In either case, it takes substantial faith to suppose such similarities provide an adequate basis for neo-Darwinism*.

Finally, the notion that unguided, purposeless processes could transform the ears of a land-dwelling, freshwater, hippo-like animal into the echolocation system* of a sea-dwelling, saltwater whale—and that in a mere matter of 10 million years—I freely confess takes more faith than I can muster.

Darwin had an excuse. In his day fossil transitions were relatively scarce. Moreover, the science of genetics had not yet evolved. As such, the hip neo-Darwinian notion that a three-thousand-pound hippo could evolve into a three-hundred-thousand-pound whale through a series of genetic missteps is no less a stretch than a Darwinian bear evolving into a whale with a tail.

For further study, see William A. Dembski and Jonathan Wells, *The Design of Life: Discovering Signs of Intelligence in Biological Systems* (Dallas: Foundation for Thought and Ethics, 2008); and Michael J. Behe, *The Edge of Evolution: The Search for the Limits of Darwinism* (New York. Free Press, 2009).

Can We Be Certain That Evolution Is a Myth?

Evolution is rightly dubbed "a fairy tale for grownups," but I think it's something much worse. I call it a cruel hoax! In fact, the arguments that support evolutionary theory are astonishingly weak.

First, the fossil record is an embarrassment to evolutionists. No verifiable transitions from one kind to another have as yet been found. Darwin had an excuse. In his day fossil finds were relatively scarce. Today, however, there is an abundance of fossils. Still, we have yet to find the projected wealth of transitions from kinds to other kinds.

Furthermore, in Darwin's day, such enormously complex structures as a human egg were thought to be quite simple—for all practical purposes, little more than a microscopic blob of gelatin. Today, we know that a fertilized human egg is among the most organized complex structures in the universe. In an age of scientific enlightenment, it is incredible to think that people are willing to maintain that something so vastly complex arose by chance. Like an egg, the human eye or, for that matter, the earth is a masterpiece of precision and design* that could not have come into existence by chance.

Finally, while chance is a blow to the theory of evolution, the laws of science are a bullet to its head. The basic laws of science—including the laws of effects and their causes, energy conservation, and entropy*—undergird the creation model for origins and undermine the evolutionary hypothesis.

While we should fight for a person's right to believe science fiction, we must resist evolving attempts to equate the "certainty" of macro-evolution* (the evolution from one species to another species) with such scientific certainties as the law of gravity.

For further study, see Hank Hanegraaff, *Fatal Flaws. What Evolutionists Don't Want You to Know* (Nashville: Thomas Nelson, 2008).

Did God Create Inherently Flawed Eyes?

Darwin used the human eye to illustrate "organs of extreme perfection and complication." To neo-Darwinists*, however, the human eye is a case of ill-designed optics. So what gives?

First, while evolutionists believe the verted retina in invertebrates (e.g., squids) is superior to the inverted retina in vertebrates (e.g., humans), precisely the opposite is true. As a case in point, the inverted retina features a highly sophisticated neurological feedback system that enhances contrast without sacrificing detail. Even the oft-mentioned blind spot is not blind in that the dual eyes of a vertebrate provide overlapping fields of vision.

Furthermore, as has been well said by historian Henry Petroski, "All design involves conflicting objectives and hence compromise." As such, *constrained optimization* invokes the art and science of compromise in a sea of conflicting objectives. For example, a theater-size screen might be optimal for watching movies—but hardly at home.

Evolutionary biologists seem fixated on individual functionalities while forgetting functional interconnectedness. Constrained optimization requires compromise for the good of the whole.

Finally, to pontificate that the eye is poorly designed is patently shortsighted. Advancements in biology have demonstrated even straightforward structures in the eye to be enormously complex. In Darwin's day, tears were thought to be quite simple. Today whole books are devoted to them. As marvelous mixtures of water, mucins, oils, and electrolytes, they not only thwart infection but moisturize the cornea and continuously cleanse the eye of pollutant particles.

Darwin was right. Human eyes are organs of extreme perfection and complication. Even now a vast number of impulses are traveling from your eyes through millions of nerve fibers transmitting information to a complex computing center in the brain called the visual cortex. Linking visual information from the eyes to motor centers in the brain is crucial to the very process of daily living. Without the coordinated development of the eye and brain in synergistic fashion, the isolated developments of either are not only meaningless but counterproductive. The eye was flawlessly designed by God to work in synergy with the entire body.

Source (and for further study), see Henry Petroski, *Invention by Design: How Engineers Get from Thought to Thing* (Cambridge, MA: Harvard Universtiy Press, 1996); also Hank Hanegraaff, *The FACE That Demonstrates the Farce of Evolution* (Nashville: Thomas Nelson, 2001).

Are Ape-Men Fictions, Frauds, and Fantasies?

As has been well said, there is no business like bone business. *Pithecanthropus erectus*, Piltdown man, and Peking man are prime exemplars.

First and perhaps best known among the ape-men icons is *Pithecanthropus erectus*. What is not as well known is that this fictional transitional form* between apes and humans is based on nothing more than a skullcap, femur, three teeth, and a fertile imagination. Darwin protégé Sir Arthur Keith pointed to *Pithecanthropus* as an example of evolving gullibility in his profession. Nonetheless, Harvard's Richard Lewontin said *Pithecanthropus erectus*, pet-named *Java man*, ought to be taught as one of the "five facts of evolution."

Furthermore, Piltdown man is a famed fraud cleverly conceived, crudely carried out—the jaw of an ape stained to match a human skull. Ironically, the aforementioned Sir Arthur Keith declared that Piltdown "represents more closely than any human form yet discovered the common ancestor from which both the Neanderthal and modern types have been derived." And the professor was not alone. Piltdown was used for forty years to dupe unsuspecting students into thinking macroevolution* is a fact.

Finally, Peking man is pure fantasy—wish giving birth to reality. Peking man was fabricated on the basis of a dusty old tooth discovered by Dr. Davidson Black as he was about to run out of funds for his evolutionary explorations. The Rockefeller Foundation awarded Black a generous grant so he could keep on digging. While Peking man evolved over time into an interesting collection of fossils, it is hardly a credible ape-to-man transition.

One would suppose that mental digestion would improve over the years. But that has hardly been the case. In 2009, *Dariwinius masillae*, affectionately nicknamed "Ida," was dubbed the "eighth wonder of the world"—the link between humans and the rest of the animal kingdom—and the most important fossil discovery in 47 million years. Currently, however, evolutionary scientists are uniformly convinced that Ida plays no role whatsoever in human evolution.

For further study, see Hank Hanegraaff, *Fatal Flaws: What Evolutionists Don't Want You to Know* (Nashville: Thomas Nelson, 2008).

Is the Mind Identical to the Brain?

Did God create us as body-soul unities, or are we merely material beings living in a material world in which reason is reduced to a conditioned reflex?

First, logically, we recognize the mind and the brain to be different in that they have dissimilar properties. A moment's reflection is sufficient to convince a thinking person that the experience of color is more than a wavelength of light. Were we merely material, such subjective characteristics of consciousness could not be experienced.

Furthermore, from a legal perspective, merely material beings could not be held accountable for a crime committed last year because physical identity changes over time. Every day we lose multiplied millions of microscopic particles such that in seven years' time virtually every part of our material anatomy changes.

Finally, in a merely material world, libertarian freedom (freedom of the will) does not exist. In such a world, everything would be fatalistically relegated to mere mechanistic material processes.

In short, logically, we recognize nonphysical aspects of humanity, such as ego; legally, we recognize a sameness of soul, which establishes personal identity over time; and libertarian freedom presupposes we are more than mere material robots.

Source (and for further study), see Gary R. Habermas and J. P. Moreland, *Beyond Death: Exploring the Evidence for Immortality* (Eugene, OR: Wipf & Stock, 2004).

Is It Possible for a Protein Molecule to Come into Existence by Chance?

Evolutionary theory concerning how the first organized form of primitive life evolved hardly corresponds to reality.

First, there is not the slightest evidence for an evolutionary sequence among the unimaginably varied cells existing on our planet.

Furthermore, no living system can rightly be called primitive with respect to any other. Consider, for example, that life at bare minimum demands no fewer than 250 different kinds of protein molecules.

Finally, giving the evolutionary process every possible concession, the probability of arranging a simple protein molecule by chance is estimated to be one chance in 10^{161} (that's a 1 followed by 161 zeros). For a frame of reference, consider the fact that there are only 10^{80} (that's a 1 followed by 80 zeros) atoms in the entire known universe.

If in time a protein molecule were eventually

formed by chance, forming a second one would be infinitely more difficult. As such, the science of statistical probability demonstrates that forming a protein molecule by random processes is not only improbable, it is impossible—and forming a cell or a chimp, beyond illustration.

The fool says in his heart, "There is no God."

Did God Use Evolution As His Method of Creation?

Under the banner of "theistic evolution," a growing number of Christians maintain that God used evolution as his method for creation. This, in my estimation, is the worst of all possibilities. It is one thing to believe in evolution; it is quite another to blame God for it.

First, the biblical account of creation specifically states that God created living creatures according to their own "kinds" (Genesis 1:24–25). As confirmed by science, the DNA for a fetus is not the DNA for a frog, and the DNA for a frog is not the DNA for a fish. Rather, the DNA of a fetus, frog, or fish is uniquely programmed for reproduction after its own kind. Thus, while Scripture and science allow for *micro*evolution* (transitions within "the kinds"), they do not allow for *macro*evolution* (amoebas evolving into apes or apes evolving into astronauts).

Furthermore, evolution is the cruelest, most inefficient system for creation imaginable. Perhaps Nobel Prize–winning evolutionist Jacques Monod put it best: "The struggle for life and elimination of the weakest is a horrible process, against which our whole modern ethic revolts." Indeed, says Monod, "I am surprised that a Christian would defend the idea that this is the process which God more or less set up in order to have evolution."

Finally, *theistic evolution* is a contradiction in terms—like the phrase *flaming snowflakes*. God can no more direct an undirected process than he can create a square circle. Yet this is precisely what theistic evolution presupposes.

Evolutionism is fighting for its very life. Rather than prop it up with theories such as theistic evolution, thinking people everywhere must be on the vanguard of demonstrating its demise.

From one man he made every nation of men, that they should inhabit the whole earth; and he determined the times set for them and the exact places where they should live. God did this so that men would seek him and perhaps reach out for him and find him, though he is not far from each one of us.

ACTS 17:26–27

For further study, see Jay W. Richards, ed., *God and Evolution: Protestants, Catholics, and Jews Explore Darwin's Challenge to Faith* (Seattle: Discovery Institute Press, 2010).

Can Aliens Account for the Origin of Life?

While *panspermia* is dispensed in different varieties, the basic notion is that life came to earth via aliens (directed) or meteorites (undirected). No less an authority than Darwinist Richard Dawkins considers the notion "an intriguing possibility." In reality, panspermia—literally, "seeds everywhere"—does little to solve the naturalistic conundrum concerning the origin of life.

First, while Dawkins is moving in the right direction by entertaining intelligent design*, he has not yet arrived at an answer for the origin of life. If life originated on earth via aliens (directed panspermia), the next question is: "How did alien life come to be?" Infinite regress* does not answer the question of source; it merely multiplies the effects.

Furthermore, even swallowing the desperate notion that life miraculously originated somewhere else in the universe, the odds that it would survive such lethal threats as ultraviolet radiation on a meteoritic journey to the surface of the earth (undirected panspermia) is effectively nonexistent.

Finally, philosophical naturalism* can no more account for the origin of life on earth than it can for the origin of life elsewhere in the universe. The biological organization of life is simply too complex for formation apart from the intentional design of an intelligent Designer.

In sum, panspermia—directed or otherwise— does not plausibly account for the origin of life.

> [He is] the Maker of heaven and earth,
> the sea, and everything in them.
>
> PSALM 146:6

For further study, see Fazale Rana and Hugh Ross, *Origins of Life: Biblical and Evolutionary Models Face Off* (Colorado Springs: NavPress, 2004).

Did Darwin Have a Deathbed Conversion?

In order to demonstrate the falsity of evolution, believers routinely communicate the account of Darwin's deathbed conversion. Evolutionists have attempted to counter them by loudly protesting that Darwin died believing that Christianity was a fraud.

First, we should note that whether Darwin did or did not renounce his theory addresses neither the truth nor the falsity of the evolutionary paradigm. Maybe Darwin renounced evolution because he was senile. Maybe he was under the influence of a mind-altering drug. Or maybe he just hedged his bets with an "eternal fire insurance policy."

Furthermore, in *The Darwin Legend*, James Moore painstakingly documents that there is *no* substantial evidence that Darwin ever renounced evolution. There is, however, abundant evidence that he consistently held his evolutionary paradigm to the death.

Finally, as followers of the One who proclaimed himself to be not only "the way" and "the life" but also "the truth" (John 14:6), *we must set the standard for the evolutionist, not vice versa.* Evolution would not be false because Darwin rejected it. Evolution is false because it does not correspond to reality.

For further study, see James F. Moore, *The Darwin Legend* (Grand Rapids: Baker, 1994).

Creation and Evolution:
Advanced Reading

Is Intelligent Design Really Science?

Richard Dawkins, professor of the Public Understanding of Science at Oxford and arguably the best-known Darwinist on the planet, claims those who do not believe in evolution are "ignorant, stupid, or insane." In place of rhetoric and emotional stereotypes, intelligent design (ID) proponents hold to reason *and* empirical* science.

First, ID proponents are willing to follow scientific evidence wherever it leads. ID theorists neither presuppose nor preclude supernatural explanations for the phenomena they encounter in an information-rich universe. As such, the ID movement rightly practices open-minded science.

Furthermore, ID begins with the common scientific principle that intelligent design is detectable wherever there is specified, organized complexity (information). This design principle is central to many scientific fields, including archaeology, forensic pathology, crime scene investigation, cryptology, and the search for extraterrestrial intelligence (SETI). When applied to information-rich

DNA, irreducibly complex biochemical systems, the Cambrian Explosion*, as well as the fact that earth is perfectly situated in the Milky Way for both life and scientific discovery, the existence of an intelligent Designer is the most plausible scientific explanation.

Finally, although its conclusions are not worldview-neutral, ID lends no more support to Christian theism than Darwinian evolution lends to atheism. Thus, the appropriateness of ID for public education ought to be judged on the basis of the theory's explanatory power, not on its metaphysical implications*.

> For since the creation of the world God's invisible qualities—his eternal power and divine nature—have been clearly seen, being understood from what has been made, so that men are without excuse.

ROMANS 1:20

For further study, see William Dembski, *The Design Revolution* (Grand Rapids: IVP, 2005). See also Francis J. Beckwith, "Intelligent Design in the Schools: Is It Constitutional?" *Christian Research Journal,* vol. 25, no. 4 (2003), available through the Christian Research Institute (CRI) at www.equip.org.

Is Evolution Racist?

While Darwin apologist Richard Dawkins has cleverly attempted to absolve his exemplar from overt racism, his slight of mind would be humorous were it not so pathetic.

First, one need only read Darwin's writings to recognize that his *Preservation of Favored Races in the Struggle for Life* hypothesis clearly involves "races," not merely "individuals within races." In *The Descent of Man*, for example, Darwin speculated, "At some future period, not very distant as measured by centuries, the civilized races of man will almost certainly exterminate and replace throughout the world the savage races." In addition, Thomas Huxley, who coined the term *agnostic* and was the man most responsible for advancing Darwinian doctrine, overtly communicated, "No rational man cognizant of the facts, believes that the average Negro is the equal, still less the superior, of the white man."

Furthermore, for evolution to succeed, it is as crucial that the unfit die as that the fittest survive. In his book *Bones of Contention*, Marvin Lubenow said it well: "If the unfit survived indefinitely, they would continue to 'infect' the fit with their less fit genes. The result is that the more fit genes would be diluted and compromised by the less fit genes, and evolution could not take place." Adolf Hitler's philosophy that Jews were subhuman and Aryans supermen led to the extermination of 6 million Jews. In the words of Sir Arthur Keith, a militant anti-Christian physical anthropologist: "The German Fuhrer, as I have consistently maintained, is an evolutionist; he has consciously sought to make the practices of Germany conform to the theory of evolution."

Finally, while the evolutionary racism of Darwin's day is currently politically incorrect, current biology textbooks still display residues of racism. For example, the inherently racist recapitulation theory (*ontogeny recapitulates phylogeny*: in the course of its development, the embryo repeats the evolutionary history of its species) is not only common fare in science curricula, but was also championed by no less a luminary than Carl Sagan. Despite the fact that molecular genetics has underscored the utter falsity of the dogma, neo-Darwinists* such as Harvard's

Stephen J. Gould continued to forward the pretext while decrying that "recapitulation provided a convenient focus for the pervasive racism of white scientists."

> God created man in his own image,
> in the image of God he created him;
> male and female he created them.

GENESIS 1:27

For further study, see Hank Hanegraaff, *Fatal Flaws: What Evolutionists Don't Want You to Know* (Nashville: Thomas Nelson, 2008).

What Is the Cambrian Explosion?

The Cambrian Explosion* is biology's version of the Big Bang. Just as cosmology's Big Bang undid the dogma of an eternal universe, biology's Big Bang uprooted Darwin's Tree of Life.

First, if all of geological history were compressed into a twenty-four-hour clock, most of the distinct animal forms the world has ever known would appear suddenly within a two-minute time span at around the twenty-first hour. The abrupt and simultaneous appearance of this wide array of complex body plans signals an infusion of a vast amount of information that can rightly be attributed only to an intelligent Designer*.

Furthermore, Darwin theorized that every organism evolved from a common ancestor as a result of natural selection* acting on random variations. The Cambrian Explosion points in precisely the opposite direction. Darwin said it best: "The distinctiveness of specific forms, and their not being blended together by innumerable transitional links—is a very obvious difficulty."

Finally, while Darwin predicted hundreds of thousands of transitional forms* leading to the fossils of the Cambrian Explosion, none actually appear. And since Darwin, the problem has only gotten worse. The fossil record has greatly expanded. Yet all known animal body plans appear in the form they possess today. In the words of Rudolf Raff, distinguished evolutionary biologist, "All of the known animal body plans seem to have appeared in the Cambrian Radiation*."

Darwin's candor is to be commended: "If it could be demonstrated that any complex organ existed which could not possibly have been formed by numerous, successive, slight modifications, my theory would absolutely break down." And this is precisely what has happened.

For further study, see *Darwin's Dilemma: The Mystery of the Cambrian Fossil Record* (DVD) (Illustra Media, 2010).

Did Humans Evolve from Hominids?

I n the television premiere of *Ape Man: The Story of Human Evolution*, the late *CBS Evening News* anchor Walter Cronkite confessed that his "father's father's father's father, going back maybe a half-million generations—about 5 million years—was an ape." Is Cronkite right? Or is this an illustration of the anti-knowledge surrounding ape-men?

First, whether in *Ape Man*, *National Geographic*, or *Time*, the ape-to-man icon has become the argument. Put another way, the illustration of a knuckle-dragging ape evolving through a series of imaginary transitional forms* (hominids*) into modern man has appeared so many times in so many places that the picture has evolved into the proof. In light of the fanfare attending recent candidates nominated by evolutionists to flesh out the icons of evolution, we would do well to remember that past candidates such as Lucy* have bestowed fame on their finders but have done little to distinguish themselves as prime exemplars in the process of human evolution.

Furthermore, as the corpus of hominid fossil specimens continues to grow, it has become increasingly evident that there is an unbridgeable chasm between hominids and humans in both composition and culture. Moreover, homologous structures (similar structures on different species) do not provide sufficient proof of genealogical relationships—common descent* is simply an evolutionary assumption used to explain the similarities. To assume that hominids and humans are closely related because both can walk upright is tantamount to saying hummingbirds and helicopters are closely related because both can fly. Indeed, the distance between an ape, which cannot read or write, and a descendant of Adam, who can compose a musical masterpiece or send a man to the moon, is the distance of infinity.

Finally, evolution cannot satisfactorily account for the genesis of life, the genetic code, or the ingenious synchronization process needed to produce life from a single fertilized human egg. Nor can evolution satisfactorily explain how physical processes can produce metaphysical realities such as consciousness and spirituality. The insatiable drive to produce a "missing link" has substituted selling, sensationalism, and subjectivism for solid science.

For further study, see Hank Hanegraaff, *The FACE That Demonstrates the Farce of Evolution* (Nashville: Thomas Nelson, 2001).

How Serious Are the Consequences of Believing in Naturalistic Evolution?

More consequences for society hinge on the cosmogenic* myth of evolution than on any other. Among them are the sovereignty of self, the sexual revolution, and survival of the fittest.

First, the supposed death of God in the nineteenth century ushered in an era in which humans proclaimed themselves sovereigns of the universe. Humanity's perception of autonomy led to sacrificing truth on the altar of subjectivism. Ethics and morals were no longer determined on the basis of objective standards but rather by the size and strength of the latest lobby group. With no enduring reference point, societal norms were reduced to mere matters of preference.

Furthermore, the evolutionary dogma saddled society with the devastating consequences of the sexual revolution. We got rid of the Almighty and in return got adultery, abortion, and AIDS. Adultery has become commonplace as evolutionary man fixates on feelings rather than fidelity. Abortion has become epidemic as people embrace expediency over ethics. And AIDS has become pandemic as people clamor for condoms apart from commitment.

Finally, evolutionism has popularized such racist clichés as *survival of the fittest.* Nowhere were the far-reaching consequences of such cosmogenic mythology more evident than in eugenics—the pseudoscience that hypothesized that the less fit genes of inferior people were corrupting the gene pool. As a result, segments of our society—including Jews and blacks—were subjected to state-sanctioned sterilization.

Thankfully, eugenics has faded into the shadowy recesses of history. The tragic consequences of the evolutionary dogma that birthed it, however, are still with us today.

> Since they did not think it worthwhile to retain the
> knowledge of God, he gave them over to a depraved mind,
> to do what ought not to be done.
>
> ROMANS 1:28

For further study, see Hank Hanegraaff, *Fatal Flaws: What Evolutionists Don't Want You to Know* (Nashville: Thomas Nelson, 2008).

Is Punctuated Equilibrium Legitimate Science?

Harvard's Stephen Jay Gould posited that "a species does not arise gradually by the steady transformation of its ancestors; rather, it appears all at once and 'fully formed.'" In other words, species remain relatively unchanged for long periods of time (equilibrium) followed by intervals of evolutionary change too rapid to be captured in the fossil record (punctuation). To say punctuated equilibrium is a leap of faith into a chasm of credulity would be to understate its deficiencies.

First, though popular, this evolutionary postulate is motivated by what Dr. Gould referred to as the "extreme rarity of transitional fossils*." Thus, it is a classic argument from silence.

Furthermore, punctuated equilibrium flies in the face of the science of genetics. The DNA for a reptile is not the DNA for a bird. Each is uniquely programmed for reproduction after its own kind.

Finally, even if the evolutionary jumps are from scales to feathers—rather than from reptiles to birds—the leaps are still too fantastic. The effects of jumping genes would not be a modern bird but a monstrosity.

Perhaps the greatest tragedy is that punctuated equilibrium is a theological paradigm pushed on children by such prestigious scientific think tanks as the American Association for the Advancement of Science, the American Council on Education, and the Association for Childhood Education International.

> If anyone causes one of these little ones who believe
> in me to sin, it would be better for him to have
> a large millstone hung around his neck and to be
> drowned in the depths of the sea.
>
> MATTHEW 18:6

For further study, see Hank Hanegraaff, *The FACE That Demonstrates the Farce of Evolution* (Nashville: Thomas Nelson, 2001).

Why Did Heliocentrism Triumph over Geocentrism?

Though the geocentric/heliocentric debate is often posited as science versus Scripture, in reality it is science against science.

First, three centuries before Christ, Aristarchus observed the size and distance of the sun and moon and projected the only luminous object in our planetary system to be at its center. As such, he rejected geocentrism in favor of heliocentrism. Despite his calculations, the views of Aristotle—who not only embraced geocentrism but the scientifically implausible notion of an eternal universe—would hold sway for almost two more millennia.

Furthermore, on the basis of the keen observation of planetary motion, Copernicus jettisoned the geocentrism canonized by Ptolemy in favor of the heliocentrism championed by Aristarchus. Tragically, the era's resolute bias toward ideal shapes prevented Copernicus from considering that planetary orbits might well be elliptical rather than circular. Not until Kepler in 1620 did observational data overcome that scientific prejudice.

Finally, a half century after Copernicus, Galileo—with telescope in hand—observed the phases of Venus and four moons of Jupiter, thus corroborating heliocentric sensibilities. Ironically, the observational data of Galileo was resisted by Roman churchmen who had canonized the geo-centrism of ancient pagan intellectuals (Ptolemy and Aristotle).

In the end, pseudoscientific explanations wedded to pseudoscriptural elucidations are a prescription for misunderstanding both.

Do Naturalists Consider Chance the Singular Cause of Evolution?

In a word—no! The more sophisticated naturalists readily admit that the chance-alone hypothesis is at best far-fetched. Thus, they posit that natural selection* or some other unintelligent nonrandom mechanism is involved in the process. This, however, hardly limits the liabilities of the evolutionary hypothesis.

First, it should be noted that there is no evidence for the suggestion that information in the genetic code can be increased through natural selection. Nor are there any known physical laws that can be invoked to account for the information-rich content of genetic material.

Furthermore, it is misleading to suggest that an accumulation of beneficial changes will produce an improved overall design. In other words, a bunch of little changes will not necessarily add up to a glorified final product.

Finally, those capable of scaling the evolutionary language barrier* immediately realize what's going on. Evolutionists speak of "natural selection" all the while pouring the meaning of intelligent design* into the words.

How long will you love delusions and seek false gods?

PSALM 4:2

For further study, see Nancy R. Pearcey, "DNA: The Message in the Message," *First Things* (June/July 1996): 13–14, online at http://www.firstthings.com/article/2007/10/002-dna-the-message-in-the-message-44 (accessed April 4, 2011).

Creation and
Re-Creation

Will Adam and Eve Receive Brand-New Bodies in Eternity?

Just as it is utterly impossible for a common caterpillar to imagine becoming a beautiful butterfly, it is impossible for saved humanity to fully comprehend what their bodies will be like in eternity. Of one thing we can be certain, neither our first parents nor we will receive brand-new bodies.

First, consider the biblical portrait of the resurrected Christ's body. As there is a one-to-one correspondence between the body of Christ that died and the body of Christ that rose, so too our resurrection bodies will be numerically identical to the bodies we now possess. In other words, the redeemed—whether Adam, Eve, or anyone else—will not receive brand-new bodies but their original bodies totally *transformed*.

Furthermore, while orthodoxy does not dictate that every cell of our present bodies will be restored in eternity, it does require continuity between our earthly bodies and our heavenly bodies. To inform our thinking, the apostle Paul provided us with a seed analogy (1 Corinthians 15:35–38). As a seed is transformed into the body it will become, so too our mortal bodies will be transformed into the immortal bodies they will be. While the blueprints for our glorified bodies are in the bodies we now possess, the blueprints pale by comparison to the buildings they will be (2 Corinthians 5:1).

Finally, while Paul spoke of a "spiritual body" (1 Corinthians 15:44), he did not intend to communicate that we will be re-created as spirit beings. Rather, our bodies in eternity will be supernatural, Spirit-dominated, and sin-free—*super*natural, not simply natural; Holy Spirit dominated, not dominated by hedonistic sensations; sin-free, not slaves to sin.

Although we continue to struggle against sinful proclivities in the present, we eagerly await a metamorphosis that will transform our natural bodies into bodies that are immortal, imperishable, and incorruptible.

Our citizenship is in heaven. And we eagerly await
a Savior from there, the Lord Jesus Christ, who, by the
power that enables him to bring everything under
his control, will transform our lowly bodies so that
they will be like his glorious body.

PHILIPPIANS 3:20–21

For further study, see Hank Hanegraaff, *Resurrection* (Nashville: Thomas Nelson, 2002).

How Can the Eternal Son of God Be the "Firstborn over All Creation"?

In his letter to the Colossians, the apostle Paul called Jesus Christ the "firstborn over all creation" (1:15). But how can Christ be both the eternal Creator of all things yet himself be the *firstborn*?

First, in referring to Christ as the *firstborn*, Paul had in mind preeminence or "prime position." This usage is firmly established in the Old Testament. For example, Ephraim is referred to as the Lord's "firstborn" (Jeremiah 31:9) even though Manasseh was born first (Genesis 41:51). Likewise, David is appointed the Lord's "firstborn, the most exalted of the kings of the earth" (Psalm 89:27) despite being the youngest of Jesse's sons (1 Samuel 16:10–13). While neither Ephraim nor David was the first one born, they were firstborn in the sense of preeminence.

Furthermore, Paul referred to Jesus as the first-born *over* all creation, not the firstborn *in* creation. As such, "He is before *all things*, and in him *all things* hold together" (Colossians 1:17). The force of Paul's language is such that Arians like the Jehovah's

Witnesses have been forced to insert the word *other* ("he is before all [*other*] things") in their New World Translation of the Bible in order to demote Christ to the status of a created being.

Finally, as the entirety of Scripture makes plain, Jesus is the eternal Creator who spoke and the limitless galaxies leaped into existence. In John 1, he is overtly called "God" (v. 1), and in Hebrews 1, he is said to be the One who "laid the foundations of the earth" (v. 10). And in the very last chapter of the Bible, Christ refers to himself as "the Alpha and the Omega, the First and the Last, the Beginning and the End" (Revelation 22:13). Indeed, the whole of Scripture precludes the possibility that Christ could be anything other than the preexistent Sovereign of the universe.

He is the image of the invisible God, the firstborn over all creation. For by him all things were created: things in heaven and on earth, visible and invisible, whether thrones or powers or rulers or authorities; all things were created by him and for him.

COLOSSIANS 1:15—16

For further study, see Robert L. Reymond, *Jesus, Divine Messiah: The New Testament Witness* (Tain, Ross-shire, Scotland: Christian Focus Publications, 2003).

Will the Created Cosmos Be Resurrected or Annihilated?

When the apostle Peter wrote, "We are looking forward to a new heaven and a new earth, the home of righteousness" (2 Peter 3:13), he was not describing an earth altogether different from the one we now inhabit but rather the cosmos resurrected without decay, disease, destruction, or death.

First, we might rightly conclude that the cosmos will be resurrected, not annihilated, on the basis of Christ's conquest over Satan. As the cross ultimately liberates us from death and disease, so too it will liberate the cosmos from destruction and decay (Romans 8:20–21).

Furthermore, the Greek word used to designate the newness of the cosmos is *kainos*, meaning "new in quality," not in kind—a cosmos existing in continuity with the present creation. Put another way, the earth will be thoroughly transformed, not totally terminated. When a flood destroys an island, it does not cease to exist, nor will the earth when it is renewed by fire.

Finally, the metaphor of childbirth is instructive: from Paradise lost will emerge Paradise restored. As Scripture puts it, "The whole creation has been groaning as in the pains of *childbirth* right up to the present time" (Romans 8:22). But, like a mother, earth will birth a new Eden in which God will wipe every tear from our eyes (Revelation 21:1–4).

> *We know that the whole creation has been groaning as in the pains of childbirth right up to the present time. Not only so, but we ourselves, who have the firstfruits of the Spirit, groan inwardly as we wait eagerly for our adoption as sons, the redemption of our bodies.*
>
> Romans 8:22–23

For further study, see Hank Hanegraaff, *Resurrection* (Nashville: Thomas Nelson, 2002).

Final Thoughts

Whilst on board the Beagle I was quite orthodox, and I remember being heartily laughed at by several of the officers (although themselves orthodox) for quoting the Bible as an unanswerable authority on some point of morality.

CHARLES DARWIN

In bringing *The Creation Answer Book* to a close, I want to impress a face on the canvas of your consciousness. Not just any face, but *the FACE that demonstrates the farce of evolution*. It has a sharply receding forehead cascading abruptly into a heavy brow ridge. Its mouth juts open, revealing ape-like teeth. Its eyes are deepset and pensive. They are the eyes of a philosopher. The picture is worth a thousand words, its message crystal clear. This monkey is man in the making. Your earliest ancestor was not Adam but an ape. Genesis is merely the beginning of an infamous fairy tale.

Why impress this picture on your mind? Because for evolutionists this *face* is often the argument. The tables, however, are easily turned. *F-A-C-E* can serve as a memorable snapshot demonstrating that macroevolution* is no longer tenable in an age of scientific enlightenment.

The *F* reminds us of *Fossil Follies.* As demonstrated in the entry "Did Hippos Evolve into Whales?" (page 160), the evolutionary hypothesis has evolved into a whale of a tale, but fossil evidence is scant. Thus, belief in the evolutionary development of physiological wonders such as blowholes, sonar, and diving mechanisms is founded in faith rather than on fossils. Darwin's candor is to be commended: *"Geology assuredly does not reveal any such finely-graduated organic chain; and this, perhaps, is the most obvious and serious objection which can be urged against the theory. The explanation lies, as I believe, in the extreme imperfection of the geological record."* One hundred and thirty years after Darwin's death, the fossil record continues to be an embarrassment to the evolutionary hypothesis.

The *A* represents *Ape-Men—Fiction, Fraud, and Fantasy.* Perhaps best-known is *Pithecanthropus erectus*—the *face* that stared back at me from the pages of my high school textbook (see "Are Ape-Men Fictions, Frauds, and Fantasies?" on page 167). Darwin protégé Sir Arthur Keith pointed to Pithecanthropus as an example of evolving gullibility in his profession. And mental digestion has not improved much over the years. In 2009, *Darwinius masillae*, affectionately nicknamed "Ida," was dubbed the "eighth wonder of the world"—the link between humans and the rest of the animal kingdom—and the most important fossil discovery in 47 million years. Currently, however, evolutionary scientists have been forced to confess that Ida plays no role whatsoever in human evolution.

The *C* stands for *Chance*. Chance in this sense refers to that which happens without cause. Thus, chance implies the absence of both design and a designer. As documented in the entry "Is Earth a Priveleged Planet?" (see page 13), earth is a planetary masterpiece of precision and design*. It is situated between two spiral arms of a flattened spiral galaxy—the Milky Way—not too close to the core to be exposed to lethal radiation, comet collisions, or light pollution that would obscure observation of the distant universe; and not so far that a privileged planet could never form or where we would not observe different kinds of nearby stars. Earth's status in the universe is surely one of privilege. To reduce it to an accident of cosmic chance is shortsighted; to recognize it as privileged, sublime.

The *E* represents *Empirical** *Science*. Go to the drama section of iTunes movies and you'll find a propaganda piece titled *Inherit the Wind*. It features a fictionalized account of the 1925 Scopes trial, in which creationists are portrayed as bigoted ignoramuses and evolutionists as benevolent intellectuals. In the end, one is left with the notion that believing in the creation model for origins is tantamount to committing intellectual suicide. In reality, the inverse is true. As documented in the entry "Who Made God?" (see page 2), simple logic dictates that the universe is not merely an illusion; it didn't spring out of nothing (nothing comes from nothing; nothing ever could); and it has not eternally existed (the empirical law of entropy* predicts a universe that eternally existed would have died an "eternity ago" of heat loss). The only empirically plausible possibility is the universe was made by an unmade Cause greater than itself.

In truth, it is not enough to demonstrate that evolution is a farce. You must also be ready to use well-reasoned answers as opportunities for sharing the good news that the God who created us also desires reconciliation and fellowship with us. While you cannot change anyone's heart—only the Creator can do that—you can prepare yourself to effectively communicate the difference between mere religion and a meaningful relationship with the Creator of the universe. The apostle Paul underscored this very point in his famed sermon on Mars Hill:

> From one man he made every nation of men, that they should inhabit the whole earth; and he determined the times set for them and the exact places where they should live. God did this so that men would seek him and perhaps reach out for him and find him, though he is not far from each one of us. (Acts 17:26–27)

Some sneered at the message. Others were saved.

Glossary

alchemy: A medieval speculative philosophy and form of chemistry largely attempting to change common metals into gold and produce an elixir of long life.

Arabah: The hot and dry elongated depression through which the Jordan River flows from the Sea of Galilee to the Dead Sea.

benighted: Being in a state of intellectual or moral darkness; ignorant.

book of nature: A reference to general revelation, which is the revelation of God through the created order (see Romans 1:18–20; Romans 2:14–15; Psalm 19:1–4); does not include God's specific plan to reconcile the world to himself through Christ (the latter being special revelation found only in the Bible).

Cambrian Explosion (aka Cambrian radiation): As reflected in the fossil record, almost all of the animal body plans that have ever existed on earth abruptly appeared within the Cambrian period, about 530 million years ago. No traces of evolutionary precursors have as yet been discovered in the fossil record. As Oxford zoologist Richard Dawkins acknowledged, "It is as though [fossils] were just planted there, without any evolutionary history." The difficulty this poses for the theory of evolution plagued Darwin himself.

common descent: The idea within biological evolutionary theory that all living things—humans, animals, plants, bacteria, etc.—descended with modification from shared ancestors. Darwin postulated that all living organisms on the earth descended from a single common ancestor. See also **Darwinian Tree of Life**.

cosmogenic: Concerning the origin and history of the universe.

Darwinian Tree of Life: In *The Origin of Species*, Charles Darwin wrote, "All the organic beings which have ever lived on this earth have descended from some one primordial form." The only illustration Darwin included in his magnum opus is what he called the "great Tree of Life" (see page 159) depicting the natural history of life from the universal common ancestor (root) to modern species. See also **common descent**.

design: see **intelligent design**.

echolocation system: A sensory system used by whales, dolphins, bats, and other animals to discern the distance and direction of objects by emitting (usually high-pitched) sounds that echo back from the object.

empirical: Based on observation or experiment.

entropy: The measure of disorder in a closed system (also known as the second law of thermodynamics). The amount of energy that can be used to do work in an isolated system always decreases through time. The fact that the entropy of the universe increases over time implies that the universe came into being in the finite past.

evolutionary language barrier: Most intellectual disciplines develop a jargon or in-house language that can inhibit comprehension by those outside the discipline, but what is peculiarly ironic and telling about the rhetoric of evolutionary biology and related scientific disciplines is the ubiquitous reference to living organisms and their constituent parts (organs, cells, etc.) in terms of design and purpose, despite the evolutionist's incessant denial that such design and purpose is not real but only apparent. As Oxford zoologist Richard Dawkins famously wrote, "Biology is the study of complicated things that give the appearance of having been designed for a purpose." Francis Crick, codiscoverer of the double-helix of DNA, wrote, "Biologists must constantly keep in mind that what they see was not designed, but rather evolved."

exegetical liability: Incongruities and disadvantages that result from misinterpreting a text.

federal headship: Refers to the teaching that God chose the first man, Adam, to represent the entire human race such that on account of Adam's sin, all of his descendants (i.e., all of us) are born spiritually dead in sin. Christ is the "second man" who redeems all fallen humans who trust in him. As the apostle Paul put it, "For as in Adam all die, so in Christ all will be made alive" (1 Corinthians 15:22; cf. Romans 5:12–21; Ephesians 2:1–5).

Fertile Crescent: A region of the Middle East that served as the seat of ancient Mesopotamian civilization. Egypt, Assyria, Babylonia, Israel, and other nations developed in this fertile land ranging from the Nile Valley to the Tigris and Euphrates rivers.

fossil transitions: Fossil sequences representing identifiable evolutionary intermediate links between different kinds of organisms, such as between land mammals and whales, between reptiles and birds, or between primates and humans. To date, no compelling transition sequences have been discovered in the fossil record.

hominids: I use this term to refer to the classification of any bipedal primate of the family Hominidae, including species within the genera *Homo* and *Australopithecus* (see **Lucy**), which in evolutionary theory are conjectured to be closely related to humans. Only *Homo sapiens* (humans) are living today.

infinite regress: The idea that there is no first cause in a series of causes.

intelligent design: Literally refers to the purposeful arrangement of parts. Intelligent design theory maintains that particular patterns or features of the universe and of living things are best explained by appeal to an intelligent cause, not an undirected, purposeless process such as **natural selection** and random mutation.

kairological: Reference to time in relation to God's purposes as opposed to a strict *chronological* rendering of time. For example, kairologically speaking, God redeemed the world through Christ's death on the cross even though a significant percentage of the redeemed lived chronologically prior to Christ's death and resurrection.

literary polemic: A counter to a point of view sustained through literature (as opposed to common speech or writing).

literary subversion: An apologetic technique used by Old Testament and New Testament writers to communicate truth about God. These ancient writers took motifs and images common to the pagan cultures of their day and imaginatively used them to retell the story from a true perspective. For example, the psalmist subverts motifs concerning the pagan storm god Baal and attributes them to Yahweh (Psalm 29). Paul subverts the story surrounding the altar to the Unknown God to exalt Christ to the Greek philosophers in the Areopagus (Acts 17:16–34).

Lucy: Nickname for a famous **hominid** fossil skeleton (roughly 40 percent complete) classified in the genus *Australopithecus*, which lived about 3.2 million years ago.

macroevolution: The dominant theory of biological evolution maintains that all living things (plants, animals, humans, etc.) have descended with modification from shared, common ancestors. Macroevolution refers to large-scale changes— such that one species transforms into another fundamentally different species exhibiting wholly different structures and functions. For example, birds are said to have evolved from dinosaurs, and whales from hippo-like land animals. This process would require the input of vast amounts of new information into the genetic code.

metanarrative: A grand, overarching story that explains all historical experience.

metaphysical implication: Something implied at the level of a larger philosophy or worldview. For example, **intelligent design** (ID) is not a religious doctrine—it is a scientific research program with

empirically testable hypotheses—but ID has profound *implications* for the truth or falsity of one's worldview. Even so, we should carefully maintain the distinction between what ID *is* and what it *implies*.

microevolution: Changes in the gene expressions of a given type of organism occur, but a completely different species is not produced. For example, through selective breeding, dogs ranging from Great Danes to Chihuahuas have been produced from ancient wild dogs, all the while remaining dogs; and bacteria can develop resistance to antibiotics, all the while remaining bacteria. This process, perhaps misnamed, does not require the input of new information because the changes are largely a function of the genetic makeup already present in the gene pool of the species.

natural selection: The central mechanism of Darwin's theory of evolution: organisms that are best adapted to their environment tend to survive in greater numbers than those less well adapted to their environment, which results in differential reproduction such that less well adapted organisms do not produce as many offspring as the best adapted organisms. So over time the heritable traits of the best adapted organisms become common and eventually dominate within the population. The source of variation on which natural selection operates is said to be, principally, random genetic mutation. In reality, natural selection and random mutation account well for **microevolution** but not for **macroevolution**.

nature red in tooth and claw: Refers to the characteristic harshness of the natural world, from pestilence and parasitism to predation within the animal kingdom to natural disasters such as tsunamis, volcanoes, and tornados.

neo-Darwinism: Reflects current evolutionary theory. In the mid-nineteenth century, Darwin postulated that all living things descended with modification from shared ancestors by means of **natural selection** operating on random variation. But he had no justified theory of heredity (how genetic traits are passed from parents to offspring) and thus no compelling explanation for the causes of variation. In the twentieth century, random mutations within the genetic code came to be understood as the principal means of variation. Although natural selection operating on random mutation accounts well for **microevolution**, it does not account well for **macroevolution**.

old-earth creationism: I use this term to refer to a range of Christian creationist views, including **progressive creationism**, that hold in common the inspiration and inerrancy of Scripture but also hold that the Bible is silent on the age of the universe and the age of the earth. Thus, for that data we must look to the book of nature. I do not categorize Darwinian theistic evolutionism (aka evolutionary creationism) within old-earth creationism.

philosophical naturalism: The worldview that nothing exists except the natural world, and all of nature is in principle entirely explainable in terms of science (physics and chemistry). This worldview is well codified in the famous aphorism of the late atheist astronomer Carl Sagan: "The Cosmos is all that is or was or ever will be."

progressive creationism: A form of **old-earth creationism** holding that God intervened in earth's history to create new forms of life progressively over geological history as manifested in the fossil record. Generally, progressive creationists deny **common descent**.

radiometric dating: Scientists observe that some varieties of atoms spontaneously change into other atoms at a predictable rate. For example, the specific form of the chemical element potassium known as potassium-40 changes into a specific form of the chemical element argon known as argon-40 (potassium-40, which has 19 protons and 21 neutrons in its nucleus, loses a proton and gains a neutron, becoming argon-40 with 18 protons and 22 neutrons). Carefully measuring the extent of this radioactive decay within geologic samples enables scientists to determine the age of some materials.

spontaneous generation: A theory once widely held that nonliving matter could directly produce living organisms.

transitional fossils, or transitional forms: See fossil transitions.

vestigial: Lingering evidence of something that once was but no longer exists.

young-earth creationism: The view that the Bible must be interpreted to teach that God created the heavens and earth in six literal twenty-four-hour days roughly six thousand years ago (no more than ten thousand years ago). The Bible is said to establish one's framework of foundational assumptions concerning the age of the earth by which the young-earther then reads the book of nature such that science is forged into a weapon to undermine standard geology and cosmology. Instead of young-earthers and old-earthers warring against each other, however, I believe that both ought to stand arm in arm together against naturalism, which is a real threat to biblical faith in our generation. (Naturalism is well-codified in the famous aphorism of the late atheist astronomer Carl Sagan: "The Cosmos is all that is or was or ever will be.")

Additional Resources

The following resources are recommended for further study of the topics in *The Creation Answer Book*:

Archer, Gleason. *Encyclopedia of Bible Difficulties* (Grand Rapids: Zondervan, 1982).

Beckwith, Francis J. "Intelligent Design in the Schools: Is It Constitutional?" *Christian Research Journal*, vol. 25, no. 4 (2003), available through the Christian Research Institute (CRI) at www.equip.org.

Behe, Michael J. *The Edge of Evolution: The Search for the Limits of Darwinism* (New York: Free Press, 2009).

Copan, Paul, and William Lane Craig. *Creation Out of Nothing: A Biblical, Philosophical, and Scientific Exploration* (Grand Rapids: Baker Academic, 2004).

Craig, William Lane. *On Guard: Defending Your Faith with Reason and Precision* (Colorado Springs: David C. Cook, 2010).

———. *Reasonable Faith: Christian Truth and Apologetics*, 3rd ed. (Wheaton, IL: Crossway, 2008).

Darwin's Dilemma: The Mystery of the Cambrian Fossil Record (DVD) (Illustra Media, 2010).

Dembski, William. *The Design Revolution* (Grand Rapids: InterVarsity Press, 2005).

———. *The End of Christianity: Finding a Good God in an Evil World* (Nashville: B&H Publishing, 2009).

Dembski, William A., and Jonathan Wells. *The Design of Life: Discovering Signs of Intelligence in Biological*

Systems (Dallas: Foundation for Thought and Ethics, 2008).

DeWeese, Garrett J., and J. P. Moreland. *Philosophy Made Slightly Less Difficult: A Beginner's Guide to Life's Big Questions* (Downers Grove, IL: InterVarsity Press, 2005).

Dorsey, David A. *The Literary Structure of the Old Testament: A Commentary on Genesis–Malachi* (Grand Rapids: Baker, 2004).

Geisler, Norman L. *Baker Encyclopedia of Christian Apologetics* (Grand Rapids: Baker, 1998).

Gonzalez, Guillermo, and Jay W. Richards. *The Privileged Planet: How Our Place in the Cosmos Is Designed for Discovery* (Washington, DC: Regnery, 2004).

Habermas, Gary R., and J. P. Moreland. *Beyond Death: Exploring the Evidence for Immortality* (Eugene, OR: Wipf and Stock, 2004).

Hagopian, David G., ed. *The Genesis Debate: Three Views on the Days of Creation* (Mission Viejo, CA: Crux Press, 2001).

Hanegraaff, Hank. *Christianity in Crisis: 21st Century* (Nashville: Thomas Nelson, 2009).

———. *Fatal Flaws: What Evolutionists Don't Want You to Know* (Nashville: Thomas Nelson, 2008).

———. "What Is Wrong with Astrology," *The Complete Bible Answer Book: Collector's Edition* (Nashville: J Countryman, 2009).

———. *Has God Spoken? Proof of the Bible's Divine Inspiration* (Nashville: Thomas Nelson, 2011).

———. *Resurrection* (Nashville: Thomas Nelson, 2002).

———. *The Apocalypse Code: Find Out What the Bible Really Says About the End Times and Why It Matters Today* (Nashville: Thomas Nelson, 2008).

———. *The FACE That Demonstrates the Farce of Evolution* (Nashville: Thomas Nelson, 2001).

Johnson, A. F. "Gap Theory," *Evangelical Dictionary of Theology*, ed. Walter E. Elwell (Grand Rapids: Baker, 1984).

Letham, Robert. *The Holy Trinity: In Scripture, History, Theology, and Worship* (Phillipsburg, NJ: P&R Publishing, 2004).

Moore, James F. *The Darwin Legend* (Grand Rapids: Baker, 1994).

Morris, Henry M. *Men of Science, Men of God: Great Scientists of the Past Who Believed the Bible* (El Cajon, CA: Master Books, 1988).

Pearcey, Nancy R. "DNA: The Message in the Message," *First Things* (June/July 1996): 13–14, online at http://www.firstthings.com/article/2007/10/002-dna-the-message-in-the-message-44.

Rana, Fazale, and Hugh Ross. *Origins of Life: Biblical and Evolutionary Models Face Off* (Colorado Springs: NavPress, 2004).

Reymond, Robert L. *Jesus, Divine Messiah: The New Testament Witness* (Tain, Ross-shire, Scotland: Christian Focus Publications, 2003).

Richards, Jay W., ed., *God and Evolution: Protestants, Catholics, and Jews Explore Darwin's Challenge to Faith* (Seattle: Discovery Institute Press, 2010).

Sire, James W. *The Universe Next Door: A Basic Worldview Catalog*, 5th ed. (Downers Grove, IL: InterVarsity Press, 2009).

Sproul, R. C. *Not a Chance: The Myth of Chance in Modern Science and Cosmology* (Grand Rapids: Baker, 1999).

Stark, Rodney. *The Victory of Reason: How Christianity Led to Freedom, Capitalism, and Western Success* (New York: Random House, 2005).

Strohmer, Charles. "Is There a Christian Zodiac, a Gospel

in the Stars?" *Christian Research Journal*, vol. 22, no. 4 (2000).

Tada, Joni Eareckson, and Steven Estes, *When God Weeps: Why Our Suffering Matters to the Almighty* (Grand Rapids: Zondervan, 1997).

Wells, Jonathan. *Icons of Evolution: Science or Myth? Why Much of What We Teach About Evolution Is Wrong* (Washington, DC: Regnery, 2000).

Young, Davis A., and Ralph F. Stearly. *The Bible, Rocks, and Time: Geological Evidence for the Age of the Earth* (Downers Grove, IL: IVP Academic, 2008).

Hank Hanegraaff is president of the Christian Research Institute and host of the *Bible Answer Man* broadcast heard daily throughout the United States and Canada via radio, satellite radio Sirius-XM 131, and the Internet. For a list of stations airing the *Bible Answer Man*, or to listen online, log on to equip.org. Hank and his wife, Kathy, live in Charlotte, North Carolina, and are parents to twelve children.

A Selection of Books by Hank Hanegraaff

Nonfiction

Has God Spoken? Proof of the Bible's Divine Inspiration

The Apocalypse Code: Find Out What the Bible Really Says About the End Times and Why It Matters Today

The Complete Bible Answer Book: Collector's Edition

The Bible Answer Book for Students

Christianity in Crisis: 21st Century

Counterfeit Revival

The Legacy Study Bible

The FACE that Demonstrates the Farce of Evolution

Fatal Flaws: What Evolutionists Don't Want You to Know

The Prayer of Jesus: Secrets of Real Intimacy with God

The Covering: God's Plan to Protect You from Evil

The Covering: Student Edition (coauthored with Jay Strack)

Resurrection

Fiction (coauthored with Sigmund Brouwer)

The Last Disciple
The Last Sacrifice
The Last Temple

www.equip.org   (888) 7000-CRI